ANATOMY AND DISSECTION
OF THE HONEYBEE

by

H. A. DADE

London

Published by the Bee Research Association

1962

Printed by Multiprint Limited, Hull, England.

To

ROSINA E. CLARK

FOREWORD

TEN years ago I received from Mr. H. A. Dade the manuscript of a book which the Bee Research Association was to publish: *The Dissection of the Honeybee*. But some of the diagrams still had to be drawn, and in each of the years that followed there was no time in Mr. Dade's busy life to complete them. After doing research work on tropical crop diseases in the Gold Coast (where his interest in bees was first aroused), he had become Assistant Director of the Commonwealth Mycological Institute at Kew. There, among many other tasks, he built up the well-known Culture Collection.

So the manuscript remained in storage; but I refused to give up the idea of publication, because there was such a need for the book. Dr. Snodgrass's book *The Anatomy of the Honeybee* (the second edition of which appeared in 1956) is strong meat for those without a formal training in zoology, and there has been no adequate up-to-date book for others—beekeepers, amateur naturalists and younger students. And the lectures and demonstration classes which the author mentions in the Introduction had awakened a very wide interest in bee anatomy. For Mr. Dade is a superb teacher, and those who attended his extra-mural classes in the University of London and elsewhere still talk about them with enthusiasm. Practical work was a special feature of them—as it is of this book—reflecting Mr. Dade's interest in microscopy, on which he has published many papers. He has been President of the Quekett Microscopical Club three times and is now the editor of its journal; he is also a Fellow of the Royal Microscopical Society.

In September 1960 Mr. Dade retired, and by December I felt that the time was ripe for reopening the question of *The Dissection of the Honeybee*. This led not only to the completion of the diagrams but to a reassessment of the whole project. This book is the result; the reasons for including a study of anatomy as well as dissection are given in the Introduction.

The book supplies a long-felt want as a guide to practical work, with diagrams designed as a laboratory aid, and with the necessary systematic anatomical background. It will be of immeasurable value to those studying for the National Beekeeping Diploma and for the examinations of the British and Scottish Beekeepers' Associations. It will also be welcomed by many other classes of students, and by teachers in schools and colleges. The clear style in which the author's thorough and wide knowledge of the subject is presented, and his first-class diagrams, will make the book of special value to them and to many others. The author remarks in the first chapter that the interest and beauty of the complicated mechanisms of the honeybee make it relatively easy to learn about them. This may be so, to someone with specialized training: Mr. Dade's contribution—by reason of his great gifts as teacher, writer and draughtsman—is to make it true for a very much wider circle of readers.

Eva Crane,
Director, Bee Research Association.

CONTENTS

THE PLATES

INTRODUCTION

PROFESSIONAL biologists and advanced students have background knowledge, practical facilities, and a repertoire of skills and techniques. They will need no guidance other than R. E. Snodgrass's classical monograph on the anatomy of the honeybee. The situation of beekeepers, amateur naturalists, and young students is very different. They commonly have little knowledge of the relevant aspects of entomology, usually very limited facilities for practical work, and no experience of refined techniques. This book has been designed for their use.

The study of anatomy is essentially a practical exercise, for a sound conception of the organization of an animal's body and of the many wonderful structures which it contains can only be acquired by close examination and dissection. Reading about it is not enough.

The ideal way of studying anatomy might seem to be to learn as we examine and dissect, and at one time the writer contemplated a book on these lines; but in practice this is not a successful method because by this means it is impossible to learn systematically. As we 'unpack' the contents of the body we come across a confused assortment of unrelated bits and pieces, small components of different systems with different functions. It is better, therefore, first to acquire an idea, in outline, of the general lay-out of the bee's body, and then to tackle the practical work, referring back, as the work proceeds, to a systematic text. To that end, this book has been planned to combine, in one cover, the two aspects of study. Part I consists of a logically arranged description of the bee's anatomy, while Part II contains instructions for the practical work, again in a systematic order which is necessarily different from that of Part I. Part II is in fact a guide book, and the plates are topographical charts to be used as an explorer uses maps to help him to find his way in unfamiliar country. They are arranged to unfold, so that they remain in sight while pages are turned. They are semi-diagrammatic, *i.e.*, they omit details which are irrelevant at this level of work. (Photographs are quite unsuitable for our purpose.) At any time during his work, the student can turn back to Part I to refer to descriptions of organs and systems.

The chapters of Part I are arranged in a natural sequence, so far as this is possible. First there is a sketch of the evolutionary back-

1

ground which clarifies what would, in its absence, be a bewildering mass of puzzling complexity. Examination of the external features of the body is the natural and most convenient starting point, for it reflects the animal's relationship with its environment, which has determined so much of its structure. The physiological story, based on the internal anatomy, begins with food and its processing in the alimentary canal, proceeds to its transportation to the tissues by the circulating blood and its machinery, and the elimination of waste. Next comes respiration and its apparatus, supplying the oxygen needed for metabolism, the chemical conversion of food to tissues, and taking away the waste gases. The whole of the bee's activities, both physical and physiological, are controlled and co-ordinated by the nervous system, informed by the sense organs. All these functions lead to one end, the great aim of Nature: reproduction, served by its own remarkable organs; this is both the end of the sequence and the beginning of another. For the purposes of this book, we assume that the chicken preceded the egg, and the final chapter of Part I deals with the anatomy of the juvenile forms, from egg to pupa.

Much anatomical and histological detail has been omitted, because it is not essential at this level and would tend to throw the really important facts out of perspective. Thus while it is necessary to understand the general plans of the nervous, muscular, and tracheal systems, and some of their more important parts, a full knowledge of their extremely intricate details is by no means essential. Many points of detail of the drone's endophallus have little or no known significance, and are not mentioned here. On the other hand, the bee's proboscis and sting apparatus, though somewhat complicated, are important and remarkably efficient mechanisms which are clearly understandable and of fascinating interest, so they are fully described.

Many have been deterred from undertaking anatomical study by the idea that an expensive microscope is required, and by the supposed difficulty of learning how to set about the work (some have tried to cut up bees in the dry state, with no success!). It is, of course, an advantage to have a good dissecting microscope, but an entirely adequate instrument can be improvised at slight cost. This and other constructions are described in an appendix. In another appendix, simple methods of preparing useful, permanent microscope slides are given. Finally, for those to whom technical terminology based on words derived from Latin and Greek is a mystery, there is a list of the terms, their plurals, and their derivations, which indicate why these words were chosen as international labels. Their etymology not only gives these words an interest of their own, but also helps us to remember their meanings.

Thanks are due to Mr. Gilbert Nixon and Arrow Books Ltd. for permission to reproduce in the frontispiece three figures from *The World of Bees*. For accuracy and beauty, these drawings by Mr. Arthur Smith could hardly be surpassed.

The bare statement of dedication at the beginning needs amplification. Mrs. R. E. Clark, N.D.B., Surrey County Beekeeping Adviser, is really the progenitor of this book. Nearly twenty years ago, as the energetic secretary of the Twickenham and Thames Valley Beekeepers' Association, with foresighted conceptions of apicultural education, she induced the writer to embark on courses of lectures, demonstrations, and practical instruction, first in the local association, later for Middlesex County, and finally for several years in the University of London. This book has grown out of the notes and methods used, and the experience gained, in that work.

H. A. DADE.

Richmond, Surrey, 1962.

PART I

OUTLINE OF ANATOMY

CHAPTER 1

THE BACKGROUND OF EVOLUTION

Structure and Function

Every animal body contains a great many structures adapted for a variety of purposes, and in fact we cannot think about a structure without also thinking of its function. Even in a small creature like the honeybee there are so many complicated mechanisms that learning them may at first seem to be a formidable task; but their interest and beauty make it relatively easy. It is not difficult to discover the function of such organs as limbs and mouthparts by careful observation, but the precise workings of the internal organs are often obscure. In the honeybee many physiological mysteries remain unsolved. In any case, in a book of this kind, we can discuss physiology only in general terms, without going into such biochemical details as are known.

Uninformed speculation, such as fills the correspondence columns of the popular bee journals, is worthless. It is most commonly the result of thinking of the bee as a unique phenomenon, ignoring the background of what is known about other insects, and indeed about the whole animal kingdom.

The Biological Background

This background shows us the bee in correct perspective and gives us valid clues to the solution of anatomical problems. In the course of evolution, all animals have descended from much simpler forms. A few have at some time during the process reverted to less complicated forms, either almost completely, like several successful parasites, or in some details of their anatomy, even to the point of the disappearance of some organs.

In general, however, the process has been one of elaboration and specialization, and the honeybee, one of the highest insects, is one of the most specialized.

There are three sources of information about evolution: embryology, comparative anatomy, and the fossil record. The last-

5

named is not of much help in the case of the bee, though we know that the Hymenoptera, the Order of insects to which the bee belongs, made its first appearance about 150 millions of years ago. Honeybees (but not our species) are found in Prussian shales fifteen millions of years old, coeval with our first ape-men ancestors, but *Apis mellifera* has only a very incomplete fossil record, the earliest fossils being in East African gum copal only ten thousand years old; it may be a much older species. It is a matter of great interest that the evolution of the flowering plants, the most advanced and dominant group of the vegetable kindom, evolved in step with the Hymenoptera: as these insects throve on their collection of pollen for food, the plants developed a wonderful variety of mechanisms for securing pollination, and conspicuous flowers and nectar to attract the useful visitors.

The other sources of information are much more productive. As it develops from the egg-cell to the fully formed adult, an embryo passes through stages which successively resemble the juvenile ancestral forms. Thus we have learned that insects have descended from worm-like animals, and how insects' characteristic bodies and organs have been derived from the simple ancestral patterns by modification, elaboration, and specialization. But this is not enough. We must look for evidence of more recent changes elsewhere. We must compare the bee's organs and their development with those of other insects, both near relations and very distant ones. The bee's remarkable and highly specialized proboscis would be a complete mystery if we knew nothing about the mouthparts of the wasp and those of much more primitive insects such as the cockroach. In the same way, we should never be able to guess that the bee's sting is in fact a greatly modified ovipositor (instrument for depositing eggs) if we did not know that the ovipositors of other Hymenoptera, such as the sawflies and ichneumons, develop from precisely the same embryonic structures.

It is interesting and useful to consider briefly the early stages of evolution, beginning at the level of the single-celled organisms like the free-swimming algae and protozoa which swarm in ponds and ditches. These are by no means so simple as they appear at first sight, but they provide a convenient starting point. (We have only very recently begun to form ideas of still earlier evolution, from the study of viruses and the molecular chemistry of chromosomes.) The single-celled organisms are completely self-contained, each cell being able to seek favourable conditions and find food by means of physical and chemical detectors (the simple forerunners of the sense organs of higher forms), and each being able to reproduce itself and/or mate with other cells of opposite polarity. Reproduction is carried out by the division of the cell into two daughter cells, the separation of these producing two similar independent individuals. There came a time when in some cases the two daughter cells failed to separate, but remained attached to one another; apart from this change, the two individual cells continued to forage and to reproduce as before. Through continued division a small cluster of cells accumulated—a *colony*. The colonial habit, thus established, became a fixed characteristic. A number of freshwater organisms are still in this stage of evolution. Like all kinds of co-operation between individuals, the new habit was an advantage. Through connecting strands of protoplasm the units of the colony shared their communal food harvest. In the meantime, all the units of the colony continued to reproduce independently.

In the next stage of evolution, differentiation of form and function set in: some units became specialized for reproduction, while the others ceased to reproduce, their activities being restricted to collecting food. This was a profound, fundamental change; the units were no longer independent, for the food-collectors could not reproduce, while the reproductive cells could not collect their own food. The community of cells was no longer a colony of equal units—it had become an *individual*, a *multicellular* individual.

The theme of differentiation followed by transference of individuality from units to the whole was continued in the next phase of further integration. Colonies of multicellular individuals, adhering in chains (the most successful of several plans which were tried in the previous phase), were produced; then some of the multicellular units retained their power of reproduction while the remainder lost it. Again the colony was transformed into an individual, this time a *segmented* animal. The worms are still in this stage of evolution. Their shape and relatively large size have introduced new conditions, adaption to which has involved new kinds of differentiation. The worm still reveals its evolutionary history in its conspicuous segmentation—it is a long chain of similar compartments, each one of which originally contained a complete set of organs. The segments now carry pairs of appendages, oar-like swimming organs in aquatic worms, sometimes bristles in terrestrial worms like the earthworm. Movement from place to place is in the direction of the length of the body; any other method would obviously be too clumsy. So one end of the worm, the front end, comes first into contact with food, or unfavourable conditions, or enemies. Accordingly, the organs of detection have become more sensitive in this region of the body, and are becoming more elaborate structures.

Thus the more important sense organs, the eyes and antennae, begin to evolve. The mouth is also situated here. The animal is now comparatively a large one, and its segments are in close contact only with their immediate neighbours; but they must all conform to the movements of the body, and therefore some kind of apparatus for co-ordinating their actions has become necessary: thus the nervous system has come into existence, in the form of a long strand of special tissue which conducts electrical messages along the body, while knots of nervous tissue in each segment control the segmental muscles. A larger knot, or ganglion, lies near to the detectors at the front end of the body, to receive and translate the information which they give, and to pass back along the body the appropriate instructions to the segmental ganglia. This large ganglion is the germ of the brain in higher forms. Farther back, the heart is a series of segmental units, linked together, but developed only in a small number of segments; the alimentary canal is a long tube, ending in the anus at the rear and having a crop of some kind towards its other end. Excretory apparatus is complete in each of the less specialized segments.

The tail end of most worms comprises many relatively unspecialized segments, where the primitive local control still holds very largely. If the rear part of an earthworm is cut off, the animal is not much affected, and the lost segments are replaced. In this respect the worm behaves as colonies do. But if the front end of the worm is cut or damaged, the animal dies.

The tapeworm, in its fully developed state, is a curious mixture of primitive and highly specialized features. The front end of the animal carries a ring of recurved spines which anchor the worm to the host's intestine. The remainder of the body is composed of identical segments, each with a complete set of male and female reproductive organs and nothing else except a very simple twin nerve cord. All other organs have become superfluous, since the worm lives on predigested food and is quite inactive. The segments at the tail end become detached and pass out with the host's faeces, and are constantly replaced by new segments produced by budding behind the 'head'. Simplicity of structure here is the result of loss of organs; it is not a primitive characteristic, but adaptation to a changed habit. In the same way, structures which no longer serve a useful purpose are very often lost in the course of evolution in all kinds of animals. On the other hand, they may persist as vestigial structures (providing their retention is not a disadvantage), so it is not uncommon to find structures, sometimes quite well developed and even striking, which have no function at all.

With the information which we have, it is now possible to make a sketch (not a portrait, but a diagram) of the worm-ancestor of insects (Fig. 1). This sketch shows that the head is made up of the first six segments, the paired appendages of the second giving rise to the antennae, while those of the 4th to 6th become mouthparts.

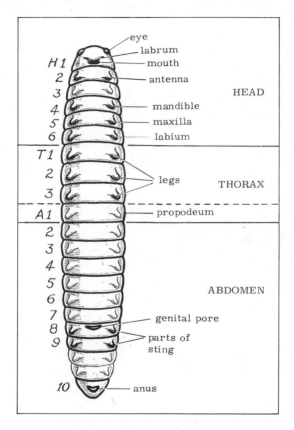

Fig. 1. Diagram of the worm-like ancestor of the insects. Parts eliminated in the bee are drawn in thin lines. Segment numbers are those of the adult bee.

The third segmental appendages do not develop (in crustacea they provide a second pair of antennae). The eyes belong to the first segment. The three segments behind the head constitute the thorax, and their appendages have become the six legs of adult insects. The remaining segments are those of the abdomen. In primitive unspecialized insects, like the bristle-tails (Fig. 2), segmentation is conspicuous in the thorax and abdomen, and it can be traced in the head also.

In nearly all insects the abdomen is separated from the three thoracic segments by a narrow waist or stalk, the *petiole* (or peduncle). In the Hymenoptera, however, the constriction of the petiole forms *behind* instead of before the first abdominal segment, which in this

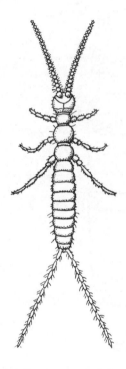

Fig. 2. *Campodea*, a bristle-tail. An unspecialized insect in which segmentation is still largely unobscured.

Order is added to the three true thoracic segments. Its structure is typical of an abdominal segment, and to conform with the standard entomological terminology it continues to be called the first abdominal, but with a special name, the 'propodeum' (meaning 'in front of the stalk or waist'). The appendages of the abdominal segments have disappeared in the bee, except those of the 8th and 9th, which have become parts of the sting in female bees and the 'claspers' in drones. In some insects the abdominal appendages are functional in larvae, *e.g.*, in the caterpillars of Lepidoptera. The 8th and 9th segments are much reduced in the bee, and telescoped inside the 7th in the adult. The next two segments disappear before the egg hatches, and the last one, now numbered as the 10th, though complete in the larva, is reduced to a small cup, pierced by the anus, in the adult bee.

In the honeybee larva (Plate 18), segmentation is conspicuous in all parts of the body except the head, where the initials of antennae and the optic lobes of the compound eyes are already laid down, as well as the mouthparts. No functional legs exist, these having been suppressed in the larva long ago, as the present immobile habit developed, but rudimentary legs persist.

In the text, as well as in the illustrations of this book, abbreviations have been avoided as much as possible, for when there are many, and their meaning is not immediately obvious, they can be very irritating. It is, however, convenient to use letters and numbers to indicate the segments, thus: H1 to H6 for the head segments, T1 to T3 for the true thoracic segments, and A1 to A10 for the abdominal segments.

In the adult bee, or the 'imago' in normal entomological terminology, segmentation is still clearly to be seen in the abdomen. In the head great specialization has obscured segmentation completely, and in the thorax it can still be traced in the sutures between the chitinized plates, though these are fused together to form a nearly rigid box.

Our sketch of the worm-ancestor, of its evolution and later extensive modification, gives us a conception of the honeybee which we should otherwise lack completely, and which, by making clear the meaning of so many otherwise inexplicable structures, greatly assists us in our study of anatomy. The story, however, is not yet finished.

The Evolution of Social Anatomy

If our conception of the bee is to be complete, we must take into account still further stages of cyclic evolution. We have discussed two cycles of integration, each followed by phases of differentiation and the resulting transference of individuality: unicellular organisms to colonies, colonies to multicellular individuals; multicellular individuals to chain-colonies, chain-colonies to segmented animals. Obviously the theme of physical integration cannot proceed any further: we cannot imagine a chain colony of insects!—for their key advantage of great activity and rapid movement would at once be lost. Nature has, in fact, made her experiments in this direction. The salps, small and primitive marine chordates in the segmented stage, pass one part of their life cycle as a long chain colony, a sort of ribbon which swims in the sea by undulatory movements. In some chalcids (small parasitic Hymenoptera), each egg divides to form a chain of embryos, which eventually breaks up, another survival of the ancient theme. How many similar experiments were tried at this level we do not know, for most of them were doomed to failure. Nevertheless, the theme *has* continued. Integration of segmented animals into colonies has occurred when the bonds between the units have been not physical, but psychic.

Examples of all stages in this process are to be found in the Hymenoptera, among the ants, bees and wasps, though it has also occurred in the unrelated termites. The females of solitary bees lay their eggs in shelters dug in the soil or contrived in tunnels in timber, etc., and in these shelters they provide each egg with quantities of pollen sufficient to feed the larvae up to pupation. The female never sees her offspring, for she dies when she has laid all her eggs. A new habit evolved when the female insect began to feed her larvae day by day. This change was a critical one, with far-reaching results: the association of parent with offspring was the beginning of the social habit. In the next stage the first batch of young insects emerge from their pupae while their mother is still with them; these are all sterile females, and they co-operate with their mother in feeding their younger sisters, as in the bumble bees. After a sufficiently strong family is built up, perfect females and males are raised, and mate. With the onset of winter the old queen and workers die, whereas the young fertilized queens hibernate, surviving to start new nests in the following summer. This plan has been improved by the honeybees by the establishment of winter stores of food, which enable the whole community to survive the winter and to make an early start in the following spring with a strong foraging force. Differentiation of function is complete: the queen does nothing but lay eggs, the workers take no part in reproduction (though, as a vestige of their former sex, they can lay unfertilized eggs when the community has lost its queen). The queen and drones cannot live without workers to feed them, and the workers cannot reproduce themselves.

In the preceding cycles, this loss of complete independence has involved the transference of individuality to the whole assembly, and thus we are faced with the strong suggestion that the honeybee community is no longer a colony, but a *community-individual*, of which the queen and drones are the reproductive organs. The keen observation of experienced practical beekeepers has led them to the concept of 'the hive mind' to account for the concerted actions and 'decisions' which are characteristic of bee behaviour. Fanciful attempts to explain the total subservience of the unit insects to the community by supposing the bees to be conscious, 'intelligent' creatures, able to confer and agree on courses of action, have no logical foundation. No such ability is possible in animals provided with the very small scrap of nervous tissue in the bee's head which can be called a "brain". There is no evidence to show that bees are not completely unconscious automata, whose behaviour is governed by reflexes. If they could think, it is unlikely that they could be so easily deceived and exploited as they are in the ordinary course of apiary management.

The highly specialized anatomy of the castes is thus associated with their social habit, and it may perhaps be said that the latter has a claim to inclusion in the study of anatomy.

Mass-reproduction by swarming is obviously wholly appropriate. The offspring is a new community-individual. But which is the offspring, the swarm or the occupant of the old nest? Normally the swarm is composed of the old queen and a majority of older workers, while the bees left in the hive are mainly a young queen, a lot of young workers, and a large batch of brood. It looks very much as though the swarm is really the *parent*, and this suggestion is strongly supported by another evolutionary theme, the progressively increasing care and endowment of offspring as animals evolve into higher forms. In the vertebrates, at the lowest level, the fishes drop enormous numbers of small eggs in the water and leave them to their fate. Birds lay only a few eggs, with large yolks (stored foodstuff), and feed and take care of their fledglings until they can look after themselves, though rather inefficiently. Mammals tend to produce still smaller numbers of offspring, and take better care of them. At the top of the scale, man produces very few children, and then feeds and guards them carefully, protecting them until they are far beyond the age of biological maturity. Among the insects, the honeybees are at the top of the scale. The swarm-parent flies away on the precarious venture of finding a new nest and making a fresh start, leaving the offspring community richly endowed with an established nest, stocks of food, and a strong force of young workers to care for the brood.

Names and related insects

The correct, official name of our honeybee is *Apis mellifera* L., which means 'the honey-bearing bee', not a very good descriptive name. It is the first name given to her by Linnaeus, and it appears in the 10th edition of his *Systema Naturae* (1758). Later, he changed the name to *Apis mellifica*, 'the honey-making bee', which is much more appropriate. Unfortunately, this name has been invalidated by the International Rules of Nomenclature, in which 1758 is chosen as the date after which properly applied names cannot be changed. Linnaeus had his second thoughts too late from this point of view.

The genus *Apis* comprises four species of the true honeybees, *i.e.*, those which store considerable quantities of honey. They are all closely related, their anatomy being very similar indeed. These species are *A. dorsata*, the giant honeybee, *A. florea*, the little honeybee, *A. indica*, the Eastern honeybee, and *A. mellifera*, the Western honeybee. The first three of these occur wild in southern Asia, and not elsewhere, whereas *A. mellifera* is indigenous only in Europe and Africa, and not elsewhere, though this species has, of course, been introduced into almost every country in the world from Europe. A variety of *A. mellifera, unicolor,* occurs in Africa; the numerous European strains (the "old English black", Italian, Cyprian, Caucasian, etc.) are only races, differing in minor characters, mainly coloration.

Only *A. mellifera* and *A. indica* can be kept in hives.

Some of the remarkable structures found in honeybees are not their exclusive property, but are shared by other related genera. The bumble bees, for example, have the same kind of proboscis, but a much longer one, which enables them to exploit flowers in which nectar cannot be reached by the honeybee; they have the same pollen-packing apparatus on their hind legs; and their sting is similar, too. The single spine on the tibia of the middle leg of the bee is often the subject of discussion; but it is only a relic—the bumble bees have spines on all their tibiae, including those of the hind leg. Students of anatomy should examine other insects as well as the honeybee, to improve their perspective.

CHAPTER 2

EXTERNAL ANATOMY

The exoskeleton

The insects are the largest and most successful Class of the arthropods, all of which are characterized by the possession of an external skeleton (*exoskeleton*). By contrast, the vertebrates, also descended from a worm-like ancestor, have developed an internal skeleton, a more efficient structural plan, lighter in weight and permitting growth on a larger scale. The exoskeleton is, however, a satisfactory form of framework for smaller animals. By providing rigid anchorage for powerful muscles, and a system of levers operated by the muscles, it has enabled the insects to become very active and rapid in their movements, able both to catch prey and to escape from enemies with ease.

The exoskeleton has, however, set definite limits to the size of insects, perhaps fortunately for us. The matter of limitation of size is sufficiently interesting to justify explanation, which is found in the law of squares and cubes. Let us imagine a man of twice the average stature, his other measurements being in proportion. His volume, and therefore his weight, must be proportional to the *cube* of 2, i.e., he would be 8 times the weight of the average man. The power of a muscle is proportional to its area of cross section, i.e., to the *square* of its diameter. Our giant man has muscles of twice the average diameter, so they will be 4 times as strong as those of the average man. Having 8 times the weight and 4 times the strength, he will be only half as strong and half as agile, in proportion, as the average man. If it were possible to increase the relative size of the muscles (which would also necessitate increasing the strength of the skeleton to take the extra strains), the result would again be to increase weight, and there is no solution to the problem.

We have examples of the disadvantages of great size among extinct animals. The gigantic dinosaurs became extinct because their great weight, in some cases increased by heavy protective armour, rendered them so slow and clumsy that they could not compete with smaller, more active, predators. In the case of the arthropods, the exoskeleton limits size very severely. Small size is a great advantage in terms of relative strength and agility. An ant can drag prizes many times its own weight to its nest with ease, and can perform other prodigies of strength. But the largest and heaviest insect, a beetle, is only 6 inches in length.

The exoskeleton of the bee consists of the *cuticle*, which is secreted by the cellular *epidermis* which lies beneath it. There are two layers of cuticle, the very hard *exocuticle*, composed mainly of *sclerotin* (a tanned protein) and containing some *chitin*, and the *endocuticle*, a softer layer, which is mainly chitinous. Over all is a very thin greasy, waterproof layer, the *epicuticle*. The epidermis is lined on the inner side by a tough thin *basement membrane*, to which the muscles are attached. The sense organs and the nervous system itself are also derived from the embryonic skin, and the epidermis provides the sensitive cells of the minute units (sensilla) of the sense organs.

Between the rigid plates of the exoskeleton, flexible joints consist of membrane with thin layers of cuticle.

A great deal of the area of the exoskeleton is covered with hairs, also composed of cuticle, and on the inside of the shell are ingrowths in the form of thickened plate edges, *phragmae* (fence-like ridges), or *apodemes* (peg-like extensions of thickened plate edges), which serve as strong places of attachment for muscles, and also struts to stiffen the shell where extra rigidity is required.

The waterproof nature of the cuticle prevents loss of moisture and also retains the blood, with which the body cavity is filled.

A few useful terms in anatomy must be defined. Sometimes such adjectives as 'back' and 'front' are ambiguous because we are accustomed to use them with reference to our own bodies, with their upright posture, and when they are applied to creeping animals, whose bodies are normally horizontal, confusion may result.

Ventral refers to the belly surface, the lower aspect of creeping animals, the 'front' part of man, and to any part of the body or of an organ which lies in this direction. *Dorsal* has the reverse meaning: it refers to the back, the upper side of a creeping animal, the spinal surface of man, or to any part of the body or of an organ which lies in this direction. The vertex (crown of the head) is on the dorsal aspect of an insect's head; what is it in man?

Anterior refers to the head end of the body, or to any part of the body or of an organ which lies in this direction; and *posterior* is the reverse—it refers to the tail end or to any part of the body or organ which is nearer to the tail end.

Lateral, of course, refers to a side of the animal, between the dorsal and ventral aspects.

Proximal refers to the part of an appendage of any sort which is next to the main body or organ, and *distal* refers to the other, distant end.

Thus, the foot is at the distal end of a leg, and the coxa, or hip, is at its proximal end. These terms should be used when it is not easy to indicate meaning in any other way.

The head (Plate 2)

The head is a roughly triangular and flattened box, and is composed of the front six segments, H1 to H6, though all trace of segmentation has disappeared through extensive specialization. Plate 2C shows the anterior aspect of a worker's head. The regions flagged in the plate are the *vertex* (crown), *genae* (cheeks), the *frons* (brow), and the *clypeus*. Two small pits at the edges of the clypeus mark the position of the upper ends of the *tentoria*, two struts which strengthen the head framework. The *antennae* spring from the middle of the face, and represent the appendages of H2. On the vertex are the three *ocelli* or simple eyes, and the large *compound eyes* cover almost all of the sides of the head above the genae. Hinged to the clypeus is the *labrum* or upper lip, and on each side of it, hinged to the genae, are the *mandibles*.

The posterior aspect of the head is shown at D. Here is the *occiput* and below it the *foramen*, through which the organs inside the head are connected to the thorax. Below this lies a U-shaped hollow, the *fossa*, in which the *proboscis* is slung. The pits adjoining the foramen indicate the position of the ends of the tentoria. The principal parts of the proboscis are labelled; they will be described in the next section. The lateral aspect is shown at E, and the dorsal aspect at F.

The queen's head (A) is similar to the worker's, but slightly rounder, her proboscis is shorter, and her mandibles are toothed.

The drone's head (B) is very different. The enormous compound eyes meet at the vertex, giving the face an almost circular outline. The ocelli are displaced by the compound eyes, being forced down on to the frons. The antennae are much longer and thicker, and have an extra joint. The mandibles are very small and feeble, and are toothed.

The mouthparts

The *cibarium*, or cavity of the mouth, is continuous with the *pharynx* (gullet), which lies behind it and is the dilated end of the *oesophagus*; the oesophagus enters the head through the foramen. The roof of the cibarium terminates at the epipharynx, under the labrum. The floor of the cibarium spreads out into a sheet of membrane in the fossa, where it connects the parts of the proboscis together, and forms with them a continuous, flexible, hammock-like bag. (A full description of this part of the alimentary canal is provided in Chapter 3.) This brings us to the apparatus round the mouth opening which we call the mouthparts, derived from the appendages of the six head segments.

The mouthparts above the mouth are relatively simple. The *mandibles* or jaws (Plate 3) are hinged to the genae; they are strong, spoon-shaped organs in the worker, concave and ridged on the inner side. They are used for a variety of purposes: for shovelling pollen into the mouth (but not for crushing the grains, as is sometimes said), for handling, biting, cutting and kneading wax and for building comb, for collecting and applying propolis, for feeding brood food and pollen to larvae, for dragging débris out of the hive, for grooming, and for fighting. They are derived from the appendages of H4.

An aperture at the base of each mandible connects with the duct of the mandibular gland, the secretion from which presumably runs down the adjacent groove into the hollow of the mandible (see Chapter 3). The mandibles of the queen are toothed; this is a primitive feature and has no functional significance—it is the worker's mandible that is specialized for her numerous skills. The drone's mandibles have the same unspecialized form as the queen's, but are very small and are used only for shovelling pollen into the mouth.

The *labrum* is a sclerotized flap hinged to the clypeus. On its inner surface it bears a soft pad, the *epipharynx*, shaped to fit closely against the proboscis when the latter is in use, and to make an airtight joint with it, thus permitting fluid to be sucked into the mouth. The labrum belongs to H1.

The proboscis (Plate 3)

The proboscis, made up of the combined lower mouthparts, which are derived from the appendages of H5 and H6, is a very complicated structure, and one of the most interesting. Its component parts are shown in Plate 3. They are the *maxillae* (H5), composed of the *stipites*, *galeae*, *laciniae* (large in less specialized Hymenoptera but reduced to small lobes in the bee), and the vestigial *maxillary palps*. The stipites are joined together by the transverse lorum, and the whole is slung between the *cardines*, the proximal ends of which are articulated to pegs on the inner walls of the fossa (Figs. 3 and 4). The inner members of the proboscis all belong to the *labium*, or lower lip, derived from H6: they are the *postmentum*, articulated to the middle of the lorum, the *prementum* and, joined to the prementum, the *labial palps*, the long *glossa* (tongue), and two *paraglossae*. The glossa terminates in a small rounded *labellum* (or flabellum). The whole of the apparatus carried at the end of the prementum is sometimes referred to as the *ligula*.

The glossa is a hollow tube of thin tough membrane, flattened and curled at its sides, a section through it thus having a C-shape (Fig. 4). It is stiffened by a slender rod which can be drawn backwards by muscles inside the prementum. The glossa is thickly

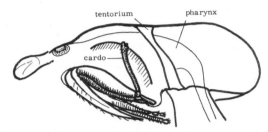

Fig. 3.   The proboscis folded, cardines swung back.

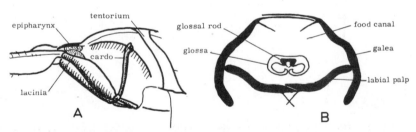

Fig. 4.   A, the proboscis extended, cardines swung forward, laciniae pressed against the epipharynx.   B, cross-section of proboscis.

covered with small hairs, and the labellum has a fringe of stouter bristles.

Saliva, the mixed secretions of the postcerebral and thoracic glands (Chapter 3), is carried by a duct to the floor of the mouth and flows into a pouch, the *salivarium* (Fig. 15), just behind the insertion of the glossa on the prementum. Here the saliva runs between the paraglossae, and then round to the underside of the glossa, where it enters the C-tube, or salivary canal, and runs down the glossa to mingle with the nectar or syrup which is being taken.

This remarkable proboscis is so highly specialized that its **origin** cannot be traced without reference to embryology and comparative anatomy.

J. A. Nelson (1917) has described the development of the mouthparts in the embryo, and three stages in this process are shown in Fig. 5. At first the embryo looks very like our sketch of the worm-ancestor, with paired appendages on most of the segments. As development proceeds, the mouth and labrum migrate to the ventral surface of the embryo, and the appendages of H4, 5, and 6 arrange themselves round the mouth. Finally the appendages of H6 coalesce, forming a single organ. This shows how the mentum of the adult

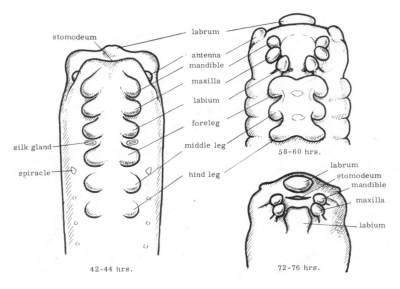

Fig. 5. Development of embryonic mouthparts, showing their progressive arrangement round the mouth, and the final fusion of the two parts of the labium (after Nelson).

bee has come to be a single and not a paired organ like the other members of the proboscis. The embryonic mouthparts are now arranged round the mouth in the positions which they occupy in the larva (Plate 18) and in the imago.

Embryology alone does not tell us the full story, however, so we must turn to comparative anatomy to complete it. The mouthparts of many unspecialized insects reveal the primitive types of the mouthparts. Fig. 6A illustrates the lower mouthparts of the cockroach, which are used to shovel or sweep into the mouth the food material which has been crumbled by powerful mandibles. The parts of the maxillae can be seen to correspond with those of the labium: the stipites with the fused mentum, the maxillary palps with the labial palps, the galeae with the paraglossae, and the laciniae with the glossae (note that the glossae are two separate organs). The wasp (Fig. 6B) shows considerable specialization for a different kind of feeding—she gnaws wood, to obtain comb-building material, with large- powerful mandibles, and sweeps this

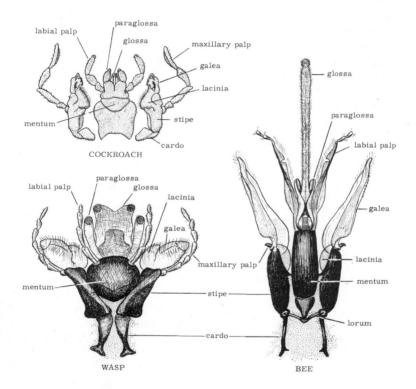

Fig. 6. Comparative anatomy of the lower mouthparts of the cockroach (*Blatta*), the wasp (*Vespa*), and the honeybee (*Apis*).

into her mouth with large laciniae and galeae; she also licks up fluid food (fruit juices, nectar—and jam!) with her tongue, composed of a broad glossa (the primitive glossae now joined together like the parts of the mentum), aided by the two paraglossae which have taken on a similar form. The two pairs of palps, bearing sense organs of taste or smell, are still well developed.

The wasp is a convenient intermediate between the primitive cockroach and the greatly specialized bee, whose mouthparts now begin to make sense (Fig. 6C). The bee's proboscis is adapted for sucking, as well as for licking. The fused glossae are now very long, and the rest of the proboscis is also lengthened, enabling the bee to reach nectaries in deep corollas. The maxillary palps are reduced to mere vestiges with no apparent function, the laciniae are also much reduced, and the galeae and the labial palps are large, important, and adapted to quite new uses. The origin of the parts of the bee's proboscis is now clear.

How does the bee use her proboscis? When it is not in use, the parts in front of the prementum and stipites are folded back and under, and the whole apparatus is swung backwards on the cardines, so that it is tucked away under the fossa (Fig. 3). When she wants to use it, the bee swings the proboscis forwards on the cardines, at the same time unfolding and extending the members carried by the stipites and prementum; they now protrude a long way in front of the mouth. The forward swing also raises the proboscis to mouth level, and there it is grasped by the mandibles, which steady it, while the laciniae are pressed against the epipharynx (the soft pad lining the labrum), which they fit perfectly, making an airtight joint. The galeae and labial palps are brought closely together, to form a tube round the glossa (Fig. 4), this being the *food canal* or sucking tube, used in the same way as we use a straw for iced drinks. Expansion of the *cibarium*, the front part of the mouth cavity, draws water or syrup up into the mouth. If the bee has to deal with a small drop of nectar, too small to suck up or too distant to reach with the sucking tube, she licks it up by dabbling the end of her glossa in the drop and then drawing the glossa up into the food canal. The fluid adhering to the glossa is thus transferred to the food canal and can then be sucked into the mouth. Solid sugar is moistened by saliva which is poured down the salivary canal of the glossa, and the sugar is scrubbed with the bristly labellum; the resulting solution is then lifted by the glossa and taken into the food canal.

These methods of feeding can be observed under the dissecting microscope by watching bees who come to syrup placed in a white saucer and left in the apiary.

The glossa can be inflated by having blood forced into it. This is done suddenly, with the object, apparently, of throwing pollen or dust off it. This sudden inflation of the glossa into a broad bladder is a remarkable sight; one can often see a shower of minute droplets or particles thrown off into the surrounding air.   An excellent drawing of the inflated glossa is given by Cheshire (1886), the English pioneer in the study of bee anatomy.

The thorax (Plates 4 and 5)

The thorax is, as we have seen, composed of four segments, the three true thoracic segments, T1 to T3, called respectively the *pro-*, *meso-* and *metathorax*, and in addition the *propodeum* (A1). All these segments are fused together to form a nearly spherical box, with the exception of small parts of the prothorax. This box is nearly rigid, but is provided with a sort of gusset which permits a slight necessary distortion.

The true thoracic segments each consist of four plates: a *tergite* (tg), or dorsal plate, a *sternite* (st), or ventral plate, and two *pleurites* (pl), or lateral plates. (These terms are preferred to 'tergum', 'sternum', and 'pleuron', used by Snodgrass, because the latter have a more general meaning, applying to the whole of the back of the animal, or its ventral surface, or side.) Some of these plates have distinct anterior and posterior parts, but we can ignore these for the present. To some extent, the boundaries of the plates are indicated by sutures, though these are not always distinct.

The tergite of the prothorax (T1) encircles the neck, like a collar. A lobe on each side projects backwards to cover the first spiracle, situated in a notch in the pleurite of the second segment. (This spiracle is the one through which the acarine mite enters, and the anatomy of the thorax in this region is important in connection with the operation for diagnosing the disease.) The pleurites of this segment are not fused to the other plates, but float on the membrane of the neck. Hooked points of these pleurites are articulated to the sides of the foramen of the head, and at their rear the pleurites are jointed to the sternite of the prothorax, which in turn is attached by membrane to the main body of the thorax. Internally, processes of the sternite form a furca partly protecting the first ganglion of the nervous system of the thorax. The forelegs, the appendages of T1, are articulated to the pleurites.

The plates of the mesothorax, T2, are all fused firmly to one another and to the tergite of T1 and the adjoining plates of T3. This tergite is the largest in the thorax, consisting of a large anterior portion, the *scutum*, strongly domed and covering most of the thorax, and behind it a transverse, prominent roll, the *scutellum*. A fissure between these parts, the *scutal fissure*, forms a sort of gusset permitting some distortion of the otherwise rigid roof, this distortion being necessary for the operation of the wings. A gap between the tergite and pleurite on each side leaves room for the complex jointing of the forewing; it is covered with tough membrane which is continuous with the membrane of the wings. Small processes and notches in the edges of the pleurites provide articulations for the wing sclerites. A large scale, the *tegula*, overlaps and protects the root of the forewing. The middle legs belong to this segment, and are articulated to the pleurites.

The plates of T3, the metathorax, are all rather small. The second spiracle, which belongs to this segment, is very small and is hidden in the membrane in a notch between the pleurites of T2 and T3. It can be seen in newly emerged bees. There is a gap for the hind wings similar to that in T2. The hind legs are articulated to the pleurites.

The fourth segment of the thorax, the *propodeum*, is by origin an abdominal segment, A1, displaced by some chance during the embryonic development of a remote ancestor of the Hymenoptera. It has the structural characteristics of the other abdominal segments, having no pleurites, but only a tergite and sternite, and similar internal and muscular equipment; its spiracle resembles the other abdominal spiracles. This spiracle is the largest in the bee, and plays an important part in respiration (Chapter 5). The tergite is very large and convex, conforming to the streamlined, spherical shape of the rest of the thorax. It nearly encircles the rear part of the thorax, and is fused to the tergite and pleurites of T3, and narrows to fit round the slender petiole. The sternite of the propodeum is a very small strap on the underside of the petiole.

The extreme specialization of the thoracic segments is the result of progressive adaptation for flight. In primitive insects, *e.g.*, dragonflies, the two pairs of wings are equally developed and of equal importance in flight; in more advanced groups the forewings have, in the course of evolution, assumed a predominant rôle. In the Hymenoptera the hind wings have become mere trailers, without driving muscles, and T3 has become correspondingly reduced in size and importance while T2 has increased in size to accommodate large and powerful flight muscles. In the Diptera (flies) the hind wings are reduced to minute vestiges, the halteres, which take no part in flight though they bear sense organs of significance in aerial navigation.

The legs (Plates 6 and 7)

The legs are the segmental appendages of T1 to T3. They all have the same organization, comprising the following *segments*, starting from the thorax: *coxa, trochanter, femur, tibia, tarsus* (with five subdivisions, the *tarsomeres*, of which the first and largest is the *basitarsus*), and the *pretarsus* or foot. These are labelled in Plate 7.

The coxae of the prothorax are articulated to the pleurites, but are connected by membrane to all the surrounding plates. Large openings between the plates allow the entrance of muscles to all the legs. Their articulations cause the three pairs of legs to move in different directions relative to the body, but in fact all the legs have extremely versatile movements, because at each joint between their segments the permitted direction of movement is different.

The feet of all the legs are similar, and may be described first (Plate 6). The pretarsus is not a tarsomere, but a segment in its own right. Its most conspicuous organ is a pair of strong recurved claws, the *ungues*, which provide secure footing on rough natural surfaces like wood, bark or leaves. Each claw has two pointed lobes, and also a stout spine. The claws are hinged to the 5th tarsomere, and are pulled down (flexed) by muscles in the femur and tibia, connected by a tendon running through the tarsus to a plate, the *unguitractor*, on the ventral side of the foot. On the dorsal surface, the *manubrium*, a plate bearing five or six long bristles, is attached to the *arcus* of the *arolium*. The arolium (sometimes erroneously called a pulvillus, which is a different organ) is normally folded and raised between the claws. When the bee reaches a smooth surface which affords no help to the claws, continued traction by the muscles pulls down the manubrium and thus spreads the arcus and unfolds the arolium, which is then pressed against the smooth surface. It is not clear how the arolium clings to the surface, for it is not provided with a sticky secretion from the *planta*, as was formerly believed.

The forelegs (Plate 6) are small and close behind the head. The hairs on the large first tarsomere, the basitarsus, are used as brushes to clear dust, other foreign matter, and pollen from the head. These legs carry the *antenna cleaners*, each consisting of a circular notch with a comb of fine hairs on the basitarsus, and a jointed spur, the *fibula*, on the tibia. When the antennae need cleaning, the foreleg is raised and passed over the antenna, which slips into the notch; the tarsus is then flexed, and the fibula closes the notch; the antenna is thus encircled by the cleaner, and is drawn through it, being brushed as it passes.

The middle leg bears no special tools. A single spine on the distal end of the tibia probably has no specific function, though it has been said to be used for detaching wax from the wax mirrors, which is incorrect, and for handling propolis as it is passed to the back legs, though this, too, is denied. The existence of the spine does not necessarily imply that it has any function at all: spines in this position, and on other segments, too, are common in insects, and are an indication of non-specialization rather than the reverse. (The bumble bee has them on all legs.) The basitarsus of this leg is somewhat flattened and is covered on the inner side with pollen brushes; it is used for clearing pollen from the thorax and passing it to the hind legs. Sometimes they fail to reach the top of T2, and some pollen remains there.

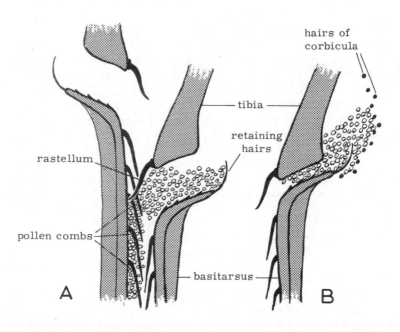

Fig. 7. Pollen packing: the diagram shows a section through the pollen press. A, press open, rastellum of right leg scraping pollen out of the pollen combs of the left leg; pollen falls on auricle. B, press closed, pollen forced through to outer side of right leg, where it is caught by the corbicula.

The hind legs (Plate 7) are highly specialized, and carry one of the most remarkable apparatuses which occur on the body of the honeybee, though it is also found in other related Hymenoptera, including the bumble bees. This is the mechanism for packing pollen and transporting it back to the nest. The inner sides of the flat, broad basitarsi are covered with rows of closely set, stiff hairs, the *pollen brushes*. These are used for brushing pollen from the abdomen after visits to flowers. The forelegs are used in a similar manner to clean the head, and the pollen they collect is moistened with honey regurgitated from the crop; the pollen, rendered sticky by the honey, is then passed back to the hind legs and received on their pollen brushes, where it mixes with dry pollen taken by the hind legs. The combined harvest collects on the basitarsi. During flight the bee rubs her hind legs together (Fig. 7), and the pollen is raked out of the pollen brushes by the *rastellum* of the opposite leg; this is a row of wide and pointed spines like the teeth of a hair comb, on the distal end of the tibia. From the rastellum the pollen falls on to the *auricle*, a sloping shelf on the end of the basitarsus; here it is retained by a fringe of hairs round the outer edge of the auricle. The *tibio-tarsal joint*, commonly called 'the pollen press' is then closed by flexing the tarsus, and the pollen on the auricle is squeezed by the opposing surface of the tibia. The rastellum, and probably also the teeth scattered on the auricle, prevent the pollen mass from escaping on the inner side of the joint, and it is forced to emerge on the outer side of the leg, where it is caught by the long, curved hairs of the *corbicula*, or pollen basket, on the tibia. The loads are then patted into compact masses by the middle legs. A single hair in the middle of the corbicula probably plays an important part in retaining the load. In the hive, the pellets of pollen are disengaged by the middle legs and dropped into cells, for packing by other workers. Dorothy Hodges, in her valuable book on the pollen loads of the honeybee (1952), gives a full and interesting account of this process.

Propolis is bitten off by the mandibles from the buds which secrete it, small pieces being taken and passed to one of the forelegs. The hind leg on the same side is thrust forwards, while the middle leg presses the propolis into the corbicula. After she returns to the hive, the collector is unloaded by another bee.

The hind legs are used for removing wax plates from the mirrors of the abdominal sternites (see under 'abdomen' in this Chapter), the plates being detached by pressing the pollen brushes against them and pushing backwards; the wax is then taken from the legs by the mandibles, after which it is kneaded by the mandibles and finally built into the comb.

The wings (Plate 6)

The wings are not derived from primitive segmental appendages, but were acquired at a much later time, after the animals had become adapted to terrestrial life. Wings appeared in many different groups of animals. In the birds the aerofoils consist of feathers, but in other animals, such as the extinct pterodactyls and the extant bats, the foils are of skin stretched between the body and greatly lengthened fingers. At the present time some animals appear to be in the early stages of wing development. The flying squirrels have sheets of loose skin between their fore and hind legs which they spread when leaping from tree to tree; they cannot fly, but are able to *glide* very efficiently. The Draco lizards undoubtedly have true wings forming, though at present they can only glide.

It is believed that the wings of insects had a similar origin as gliding organs. Fossil insects of the Carboniferous period have expansions of the prothorax closely resembling the functional wings of the other thoracic segments, but rigidly attached, and looking very much like gliding planes. Similar rigid planes occur in some recent termites.

As they form in the prepupa, the wings grow out from the thorax as small pouches containing tracheae, which disappear as the wing becomes fully formed. The 'veins' of the fully developed wing mark the position of these tracheae. They consist of stiff tubes which stiffen the thin membrane of the wing, which is also composed of cuticle. The main veins contain a little blood and a few nerve fibres. The *venation* of the wings is shown in Plate 6; it is important only in taxonomy, and needs no detailed treatment here. The wing membrane is continuous with that of the gaps between the tergites and pleurites of T2 and T3. At the root of each wing the membrane thickens into an intricate system of *articular sclerites*, the small sclerotized plates of a mechanism which, operated by muscles, imparts a variety of movements to the wing: these include not only up-and-down flapping, but also back-and-forth movements, canting of the leading and trailing edges of the wing, and furling and unfurling.

Two systems of muscles move the wings. In the more primitive insects (*e.g.*, cockroaches, dragonflies) muscles directly connected to the wing bases operate *all* movements, but in higher insects the functions of the direct flight muscles are limited. In the bee, small direct muscles attached to some of the sclerites, and at their other ends to the inner surfaces of the pleurites and coxae, serve only to adjust the cant of the wings and to furl them. The great energy required for the up-and-down strokes, which beat the air and drive the bee forward, is provided by very large *indirect flight muscles*; these are not connected to the wing bases, and they almost fill the thorax. They also operate back and forth movements of the wings on a variable axis, and the primary canting of the wings, producing pronation and supination, *i.e.*, depression of the leading edges on the down stroke, and their elevation on the upstroke.

Before we proceed to examine these muscles and the movements which they cause, it should be noted that the large indirect muscles work only on the forewings, the hind wings being trailed by the forewings and connected to them during flight by the row of *hamuli* (small hooks) on the leading edges of the hind wings, which engage automatically, at the moment of unfurling, with folds on the trailing edges of the forewings. When the wings are furled, the hooks are disengaged. Direct muscles are attached to *all* the wings (four to each forewing, three to each hind wing).

The position and attachments of the indirect flight muscles of T2

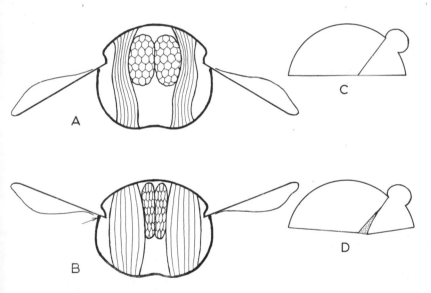

Fig. 8. Action of the indirect wing muscles. Diagrams on the left are cross sections of the thorax: A, longitudinal muscles contracted, verticals relaxed, roof of thorax raised, wings depressed; B, vertical muscles contracted, longitudinals relaxed, roof of thorax pulled down, wings raised. On the right, lateral views of the corresponding states of the thoracic tergites: C, scutal fissure closed, roof arched; D, fissure open, roof depressed.

are shown in Plate 12 (see also Plate 8). The vertical muscles, of which there are two bundles, one on each side, are attached to the domed scutum of T2 and to the sternite of this segment; the longitudinal muscles, also in two bundles, run side by side from the tergite and *1st phragma* to the *2nd phragma*. This 2nd phragma is an internal extension of the scutellum, and is a U-shaped band projecting backwards into the cavity of the propodeum. In Plate 12

c

it is seen in section. When the vertical muscles are contracted (Fig. 8), the dome of the thorax is pulled downwards towards the sternite, the resulting distortion of the otherwise rigid thorax being permitted by the scutal fissure, which opens. When the vertical muscles relax, the longitudinal muscles immediately contract, drawing the tergite towards the 2nd phragma; this causes the roof of the thorax to arch, raising the dome and closing the scutal fissure. Thus the two sets of muscles operate alternately, the thorax becoming alternately slightly flattened and slightly arched, and the wing gaps between the tergite and pleurites of T2 become correspondingly slightly narrower and slightly wider. Another effect of distortion by the vertical muscles is to stretch the slack longitudinal muscles, and *vice versa*. This is important, for the stimulus imparted by the stretching causes the stretched muscle to contract at once. Thus the alternating action of the two sets of muscles is automatically timed.

Fig. 8 shows, semidiagrammatically, the effect on the wings. The root of the wing is attached to both tergite and pleurite, the edge of the latter forming a fulcrum for the wing lever to rock on. When the tergite is pulled down, pressure is applied to the wing at a point slightly nearer to the middle of the body than the fulcrum, and the tip of the wing is quickly raised. When the tergite rises, under the influence of the longitudinal muscles, the effect on the wing is reversed, and its tip falls. Three small muscles attached to the tergite and sternite of T3 have a similar effect to that of the vertical muscles of T2, and cause the hind wing to rise; but there is no means of lowering this wing. It is drawn down by the forewing, to which it is connected by its hamuli. The enormous indirect muscles of T2, which are pink in colour and really look like meat, move the wings with great force, enabling the bee to fly strongly and to considerable distances. The wings of the drone are broader than the worker's, and his muscles are bigger; his flight is correspondingly more powerful. If a drone is held by a thumb and finger placed above and below his thorax, the vibration of the thorax can be felt as he attempts to fly. If a dead bee is held in the same way, pressure on the thorax will cause the wings to rise, and they will fall when the pressure is released.

On each side of the thorax, the anterior point of the scutellum, which projects towards the root of the forewing, bears upon one of the sclerites, the *1st axillary*. As the scutal fissure opens and closes, the point of the scutellum moves back and forth in relation to the scutum and wing base, and thus alternately pulls and pushes the 1st axillary sclerite, causing it to rock and twist the wing base. The leading edge of the wing is therefore forced downwards (pronated) on the down stroke and it is raised (supinated) on the upstroke (Fig. 8).

The tip of the wing describes a long and narrow, slanting figure of eight, the direction of movement being clockwise in the lower loop and anti-clockwise in the upper. The figure slants forwards at its base and backwards at its apex; the degree of slant can be varied, tending towards the vertical in forward flight and towards the horizontal in backward flight, an intermediate position being taken up when the bee is hovering.

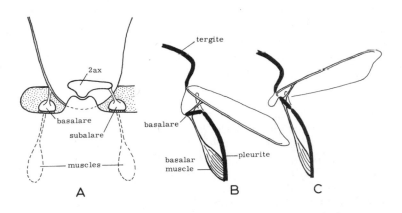

Fig. 9. Diagram, greatly simplified, showing the feathering action of the direct wing muscles. A, left forewing, seen from outside the body: the 2nd axillary sclerite (2ax) of the wing is articulated in a notch on the pleurite; the basalare and subalare are hinged to the pleurite and float on the membrane (dotted); their direct muscles are inside the thorax. B, cross-section of part of thorax: basalare pulled down by its muscle, drawing down with it the leading edge of the wing. C, basalar muscle relaxed, while the subalar muscle (hidden) pulls down the subalare and with it the trailing edge of the wing.

The remainder of the flight mechanism consists of the direct flight muscles and their attachments. Fig. 9, A, represents diagrammatically the left forewing in the raised position, as seen from outside the body. The *2nd axillary sclerite* (*2 ax*) in the root of the wing is articulated to a notch in the pleurite of T2 (this is the fulcrum mentioned above). This sclerite is also connected to a small plate, the *subalare*, which is hinged to the pleurite just behind the notch; inside the body wall this plate is connected by a tendon to the subalar muscle, the other end of which is anchored to the coxa of the second leg. In front of the notch a similar plate, the *basalare*, is also hinged to the pleurite, and is similarly worked by a muscle, the basalar muscle, anchored on the inside face of the pleurite. A similar but not quite identical arrangement exists for the hind wing. Contraction of the basalar muscle causes the basalare to swing inwards on its hinge, and it then pulls the leading edge of the wing downwards. The two extreme positions of the basalare are shown in section at B and C in Fig. 9. The opposite effect is produced by contraction of the subalar muscle, which pulls down the subalare and with it the trailing edge of the wing. These twisting adjustments of the wing can be used to modify the regular pronation and supination produced by the action of the point of the scutellum on the 1st axillary sclerite, on one side or the other, independently. By this means the bee can correct yawing, rolling, and pitching caused by air currents, these terms being used in the nautical sense; thus the bee can hold its course and maintain its equilibrium.

The wings are not merely rowing organs, like oars, but they also act as aerofoils, and modern knowledge of aerodynamics has thrown new light on the phenomena of insect flight.

The above description of the bee's flight is much simplified. We owe our knowledge of the relevant anatomy to Snodgrass (1942, 1956), but his account of the action of the mechanism has been corrected, in part, by J. W. S. Pringle in his valuable monograph (1957) on insect flight. Professor Pringle is also responsible for a very fine diagrammatic motion film, 'The wing mechanism of the honeybee', in which the very intricate anatomy and its action are clearly shown.

Furling and unfurling of the wings may be simply described as follows. In Fig. 10 the dotted lines indicate lines along which the unfurled wing can fold. In the triangular area between the folding lines is the 3rd axillary sclerite (3 *ax*). A muscle attached to the sclerite causes it to turn over, and then the wing folds along the lines, assuming the furled position. Another opposing muscle reverses the movement.

The movements of the bee's wings in flight have been studied by means of the stroboscope, and also, both visually and photographically, by gluing a scrap of gold leaf to the wing tips; in both methods the bee is held in a clamp, or otherwise prevented from taking off.

The bee's wings in flight vibrate at the rate of 200 to 250 cycles per second, and more rapidly at the highest speeds. (Corresponding figures for other insects are, for example: dragonfly 28 c./sec., house fly 330, mosquito 500-600, midge 1,000, the humming bird hawk moth 72, a long-winged wasp 160.) This rapid rate of vibration is too fast for control by any neuromuscular mechanism, and is determined automatically by the stimulation of stretching one set of indirect muscles by the contraction of the opposing set. Nervous control is limited to starting and stopping the muscles and increasing or decreasing their activity. Similar automatic control governs the

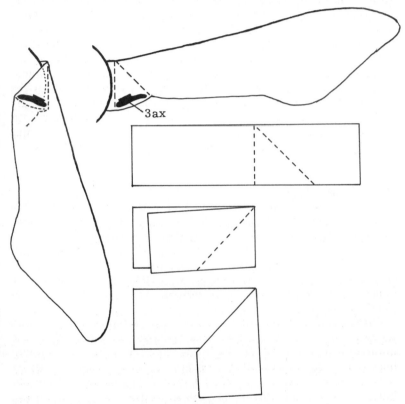

Fig. 10.   Diagram showing right forewing furled and unfurled.   3ax, 3rd axillary sclerite.   Broken lines are lines of folding.   Below, paper strip model.

heart beats of small birds (in the canary, 1,000 pulsations per minute), and the alternating contractions of the ventricles of the human heart are kept accurately out of step by a small strip of muscular tissue which connects them.

Many observers have studied the speed attained by flying bees, and from their results it appears that the average cruising speed is about 15 miles per hour, though this may be much faster, up to 25 m.p.h., for short periods. Distance is said to be covered more quickly when bees travel against the wind than when the wind is behind them, when they appear to 'take it easy'. The drone's wing is broader than the worker's, and his flight muscles are larger; his flight is proportionally more powerful. The antennae appear to act as air-speed and air-current indicators.

During flight much energy is expended, and to supply this large quantities of sugar must be consumed. The bee's blood normally contains as much as 2 per cent. of sugar; if the content falls below 1 per cent., the bee cannot fly, and if it is below 0·5 per cent., she is hardly capable of movement at all. Wigglesworth tells us that when a bee runs out and takes to the wing, her rate of consumption of fuel food goes up 50 times. In flight she uses up sugar at the rate of 10 mg. per hour, and she has a flying time, on a full stomach, of only 15 minutes and a range of 4 to 5 miles. This time and range may be extended by resting, while glycogen (p. 53) stored in the body is converted into sugar, or by consuming nectar en route.

The abdomen (Plate 4)

The front end of the abdomen is blunt, perforated by the aperture of the slender petiole; it tapers to a pointed rear extremity. With the head and thorax, the abdomen thus completes a whole which approaches the ideal streamlined form, and which may thus facilitate flight at high speeds. The abdomen consists of nine segments, A2 to A10, of which only six are visible, the remaining three being much reduced and concealed within A7 (see the description of the sting). Small parts of A8 and A9 are visible in the drone.

The abdominal segments have only two plates each, tergite and sternite; there are no pleurites. The tergites overlap one another, each rear edge overlapping the front edge of the next behind it. They are joined together by flexible *intersegmental membrane*, which permits them to move easily in relation to one another. On their inside surfaces they have thickened marginal ridges along the front edge, and a pair of hooked apodemes project from the ridge, these strengthened parts serving for the attachment of muscles. Each of the tergites of A2 to A8 has a pair of spiracles, those of A8 being concealed.

The tergites also overlap the sternites, and the sternites overlap one another in the same way as the tergites. The sternites also have thickened front edges, and each bears four apodemes, two in front and two at the sides. The sternites of the three castes differ in shape, those of the drone being broad, those of the queen being narrow and elongated from front to rear, thus conforming to the general form of their bodies.

The system of overlapping plates joined by flexible membrane permits a great variety of movement between the plates. Some of the principal muscles are shown in Fig. 11. By means of these muscles the abdomen can be lengthened or shortened, expanded or contracted in cross-section, and also curved in all directions, and twisted. Rhythmic expansion and contraction pump air in and out of the air sacs (Chapter 5), which thus act like bellows; the

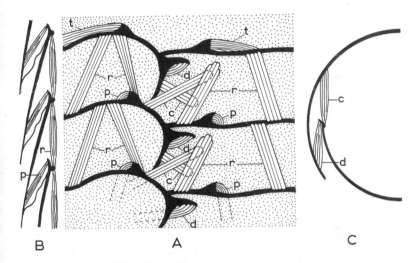

Fig. 11. The principal abdominal muscles. A, the diagram shows half of some abdominal segments, laid out; B, section through the sternites, showing how the protractor and retractor muscles act on the plates; C, half of cross-section of the abdomen, showing how the compressor and dilator muscles work. *p*, protractors; *r*, retractors; *c*, compressors; *d*, dilators; *t*, muscles in anterior segments which produce torsion.

abdomen is in a state of steady expansion (i) as faeces accumulate in the rectum, and (ii) in flying bees as nectar is collected in the crop. The abdominal muscles of the drone are very much more powerful than those of the worker, presumably to operate the sudden great contraction necessary to evert the endophallus at the moment of copulation.

Glands on the inner sides of the sternites of A4 to A7 secrete *wax*. They lie under the *'mirrors'* in the front part of these sternites (Plate 12). Scales of wax form on the mirrors, which are concealed by the preceding sternite. Very young and old bees do not secrete wax.

The *scent gland* lies under the front part of the tergite of A7, and secretes into a transverse groove, the scent canal (Plate 12). This is uncovered, by protraction of this segment, when the bee wishes to leave scent, an action which is conspicuous in fanning workers on the alighting board of the hive. (Very fine photographs of this and other activities of the bee appear in C. G. Butler's *World of the Honeybee* (1954).) Scent is used to mark food sources. It is responsible for the smell of shaken bees. It appears not to be responsible for the 'hive odour' which is recognizable by bees.

The sting (Plates 5, 10 and 11)

The sting is another of the several mechanisms in the bee's body which excite wonder and admiration. (The possession of a sting of this type is not, of course, restricted to the honeybee; many other members of the Hymenoptera have it.) It is composed of greatly modified parts of A8 and A9, and lies in the *sting chamber* inside A7. Plate 10 shows the sting *in situ* in the abdomen, as it appears in a dissection. Plate 5 shows it as it appears in a lateral dissection from the left side. This plate also shows the *spiracle plate*, all that remains of the tergite of the 8th segment; this is attached to the sting, very weakly; and the *proctiger*, all that remains of A10, now only carrying the anus, and very firmly fixed to the sting.

The sting apparatus consists of three pairs of plates, articulated together to form a system of levers which work the lancets, and accessory parts; the plates are moved by muscles, causing the lancets to dig into the victim, and pumping venom into the wound.

The paired plates are the *oblong* plates, immovably fixed on each side of the bulb and stylet of the sting, and the *quadrate* and *triangular* plates which move. Their action is shown in the diagram Fig. 12. The *protractor* muscle (198), attached to the anterior edge of the oblong plate and the posterior margin of the quadrate, pulls the latter forward when it contracts. This movement is imparted to the triangular plate, which then swings on its articulation with the oblong plate. The third corner of the plate is fixed to the proximal end of the lancet, or more accurately the *ramus* of the lancet (*r* in the figure), to which the movement is communicated. All these movements are reversed when the *retractor* muscle (199) contracts. It is attached to the posterior end of the oblong plate and to the anterior and lateral margins of the quadrate.

The ramus of the lancet is flexible and runs on a semicircular track. Thus pressure by the triangular plate pushes the ramus round

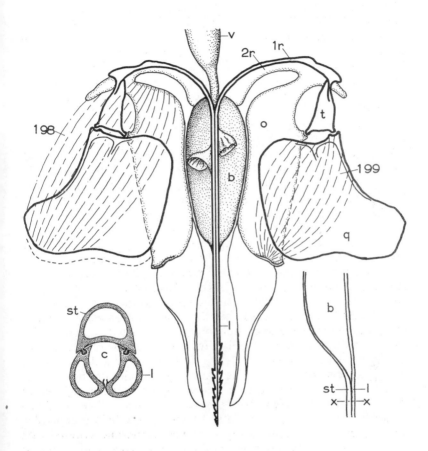

Fig. 12.   The sting mechanism: movements of the plates and lancets.   The
diagram shows the sting flattened out, the moving parts in thick lines.   On
the left side, the protractor muscle 198 contracts, pulling the quadrate plate
forward and thus causing the triangular plate to swing on its articulation
with the oblong plate; the movement is transmitted to the ramus of the
lancet (1r), which slides on the ramus of the oblong plate (2r) and causes
the lancet to be protracted.   On the right hand side the lancet is simultaneously retracted by the reverse movements produced by the contraction
of the retractor muscle 199.   The small drawings show: on the right, a
longitudinal section through part of the bulb and shaft; on the left, a
cross-section through the shaft. v, duct of venom gland; b, bulb; l, lancet;
o, oblong plate; q, quadrate plate; t, triangular plate; 1r, first ramus;
2r, second ramus; st, stylet; c, venom canal.

its track, and the lancet, which points to the rear, is caused to project in that direction. Thus the forward movement of the quadrate plate is converted to a rearward movement of the lancet. The lancets work alternately, one protracting while the other retracts.

The lancet runs freely on its track, the first part of which (the *second ramus*, *2r* in the figure) is fixed to the oblong plate. The last part of the track runs along the *stylet*. The *shaft* of the sting is composed of three parts, the stylet, a slender rigid and sharply pointed rod, and the two lancets, also very slender rods with acutely pointed tips. The stylet carries rails, reminiscent of railway lines, of circular cross-section, and corresponding grooves in the lancets fit on to the rails (Fig. 12). These rails are not continuous, but are broken at intervals, thus the friction of the system is reduced. Between the ramus and the shaft, there is an inflated organ, the *bulb* of the sting, continuous with the stylet, of which it is an extension. The track continues along the floor of the bulb. Both stylet and lancets are barbed at their points, and are thus able to take a grip on the victim's skin, and also to use this purchase to assist further penetration. At their tips the lancets separate slightly, and venom exudes through the gap. The bulb is full of venom, and this is forced into the shaft by two umbrella valves attached to the lancets inside the bulb. As its lancet is protracted, the valve opens and drives the fluid forward; on the return journey, it collapses.

It should be noted that though the stylet and lancets are hollow rods, they are closed, and do not contain or transport venom, which passes only down the canal (*d* in Fig. 12) formed by the three rods.

It is convenient to show the sting in diagrams as a flat assembly, but in fact it is not flat at all, the plates on each side curving upwards towards the dorsal aspect, and the bulb lying in the trough which they form. This is evident from the side view seen in Plate 5. When examined from above, as in Plates 10 and 11, the bulb is concealed under a semi-transparent, orange-coloured membrane, covered with short bristles; this is connected to the oblong plates, and is all that remains of the sternite of A9. The *sheath* (not a very appropriate name) is composed of two soft extensions of the oblong plates, about the same length as the shaft, which were formerly called the palps of the sting. The *furcula* is a small recurved process rising from the base of the bulb as two branches which unite after curving over it dorsally; the furcula provides attachments for muscles, the other ends of which are anchored on the oblong plates, and the whole plays an important part in preparation for stinging.

The *venom gland* ('acid gland') is a long forked tubule with slightly expanded tips; the single proximal tube widens to form a large club-shaped *sac*, in which the secretion of the gland is stored. The venom sac tapers to a narrow duct which connects with and opens into the bulb of the sting. The venom contains a mixture of enzymes (phospholipase A and hyaluromidase) and a protein (apitoxin); the enzymes release histamine and so produce the urticaria characteristic of the effect of bee stings on human victims; apitoxin has effects similar to those of cobra venom on the nervous system.

The *alkaline gland,* a whitish, strap-shaped organ which also

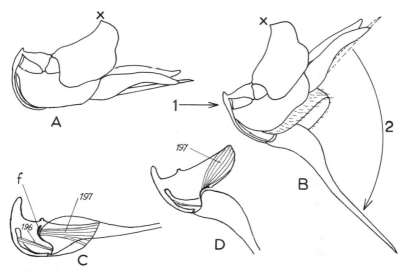

Fig. 13. Sting mechanism, lateral views, diagrammatic. A, normal position of sting: at point *x* the quadrate plate is articulated to the spiracle plate (A8, *tg*). B, the whole apparatus is swung backwards by pressure of A7, *st* at 1; at the same time the shaft is depressed, swinging through arc 2, by the furcula muscle 197.   C and D show the action of 197, which is hidden in A and B, and the opposing muscle 196. In D, 197 is contracted.

appears to supply the bulb, does in fact nothing of the sort. Its secretion is poured into the sting chamber, and its function is not known, though this has been the subject of much speculation, *e.g.,* lubricant for sting plates, an adhesive for sticking eggs to the cell floor, and so on. Another conspicuous gland of unknown function is attached to the ventral side of each quadrate plate. Its secretion has been thought to be a lubricant.

When not in use, the sting apparatus lies prone on the floor of the sting chamber, with the point of the shaft at the aperture of the 7th segment (Plate 5). In the act of stinging the bee curves her abdomen downwards, contracting the extensors of the tergites and the retractors of the sternites, and sometimes twisting her abdomen slightly with the torsion muscles of the forward segments. The sting protrudes from the 7th segment and is also canted downwards. The combined effect of these adjustments of the abdomen and the sting itself is to bring the shaft into a line almost perpendicular to the surface on which the bee clings.

Protraction of the sting apparatus is brought about by the pressure of the sternite of A7 against the base of the apparatus (Fig. 13), which swings on its attachments to the spiracle plate. Downward deflection of the shaft and bulb is the result of the contraction of the furcula muscles, which pull the furcula, and therefore the bulb and shaft, down between the oblong plates until it makes an angle of nearly 90° with its resting position. The shaft is now forced into the victim's skin by the combined force of the legs and abdominal muscles. The stabbing lancets now dig into the wound, aided by their barbs and those of the stylet.

The last ganglion in the chain (the combined ganglia of A8, 9 and 10) sends nerves to the sting apparatus and activates its muscles.

When the sting is torn out, as it is when the shaft has been embedded in human skin, or that of some other mammal, its membraneous attachments to the sting chamber and the spiracle plates, both of which are very weak, are ripped. The proctiger, however, is firmly attached to the oblong plates, and comes away with the sting, being torn off the end of the rectum. The ganglion, too, is torn away from the longitudinal commissures, and also remains attached to the sting; it continues to activate the muscles, and the lancets continue to work and to penetrate into the wound; this can easily be seen with the aid of a lens. Bees usually die within 4 or 5 days after losing their stings; young bees may last longer.

The sting of the queen, however, is firmly tied down in the sting chamber, and cannot be lost. Her sting differs from that of the worker: the shaft is curved downwards and is stouter than the worker's, and the plates are of somewhat different shape; her venom gland is very long and the sac is very large. The queen's sting is not used in depositing eggs, as has sometimes been stated: clipping off the shaft of the sting does not interfere with this operation.

The main components of the sting are derived from the segmental appendages of A8 and A9. Their initials are visible in the propupa (Plate 13), and in the lower Hymenoptera these develop into an ovipositor. The appendages of all segments, during early stages of evolution beyond the primitive worm stage, became organized into segments and branches, all on the same plan, afterwards being modified in different ways for different purposes. This can be seen more clearly in the Crustacea, which had a similar evolutionary history up to a point. Examination of the appendages of the crayfish, many of which have survived, reveals gradual differences as succeeding pairs are modified for feeding, respiration, swimming, etc. We can visualize the simple plan of each appendage as being a basal part bearing two branchlets. The basal parts of the appendages of A8 gave rise to the triangular plates, and the inner branchlets are now the lancets; the outer branchlets have disappeared. The basal parts of the appendages of A9 are the quadrate plates; the outer branchlets are the oblong plates and sheath lobes, and the inner branchlets (fused, like those of the labium) the stylet and bulb.

CHAPTER 3

FOOD AND THE ALIMENTARY CANAL

Food

The soft internal organs, or viscera, belong to eight systems, muscular, digestive, circulatory, respiratory, glandular, excretory, nervous, and reproductive. The first seven of these are all accessory to nutrition: the senses enable the bee to detect and recognize her food; the muscles, acting on the legs, wings and other organs enable her to reach her food and collect and eat it; the digestive organs and glands render the food assimilable, and the circulating blood carries the absorbed food to the tissues that need it; the respiratory system provides oxygen for the chemical processes which convert food into new tissues and energy, and removes some of the waste, while the rest of the waste is eliminated by the excretory organs. The nervous system controls and co-ordinates the whole.

An animal eats (i) to grow to maturity, and (ii) to survive, to accumulate reserves of energy for reproduction, and finally to reproduce its kind. Thus all the systems serve one end: the maintenance of the species.

Foods fall into two classes: body-building foods (proteins) and fuel foods for conversion into energy (carbohydrates and fats).

Pollen is rich in proteins, and is the only source of this food available to bees, at least in their normal economy. They possess enzymes which enable them to digest several proteins, and are able to utilize the proteins in the pollen substitutes which beekeepers sometimes provide when their bees are short of this food in the spring.

The nectar secreted by flowers is the bee's natural source of carbohydrates. Nectar consists of a solution of sugar, mainly sucrose (=cane sugar, beet sugar), of variable concentration, with traces of essential oils, which give honeys their distinctive flavours. Bees possess the enzymes necessary to digest sugars with large molecules, like sucrose and maltose; the process of inversion converts these sugars to the simpler sugars, glucose and fructose, which can be absorbed through the walls of the intestine without further alteration. Honey is the result of inversion of nectar and of concentration by evaporation of much of the water of nectar; the final product is of variable composition. It contains something like 80 per cent. of sugars, mainly glucose and fructose, and up to about

18 per cent. of water. The remaining constituents, in the neighbour-hood of 2 per cent., include very small quantities of essential oils, organic acids, minerals, and acetylcholine. Vitamins of value in human diet are present, if at all, in negligible quantities.

Sugar syrup (a strong solution of pure sucrose) is fed to stocks of bees by beekeepers, as winter stores to replace the honey which has been taken from the hives. Only pure white sugar is suitable for this purpose. Brown sugar contains indigestible substances, so does commercial glucose, which has a considerable content of dextrin. The bees themselves sometimes collect honeydew, which is secreted by aphids, on the leaves of certain trees, and this, too, contains dextrin. Starch, as well as the related dextrin, cannot normally be digested by bees, the necessary enzyme being lacking. Pollen substitutes based on leguminous flour, such as soya flour, also contain much indigestible starch. All such indigestible foods do little harm when bees are able to fly freely, but have bad effects when eaten by bees confined to the hive by winter weather (pp. 51-2).

Pollen contains starch, but it also contains diastase, and apparently this enables bees to digest pollen starch.

Fats are also fuel foods, but they do not appear to figure in the diet of bees. Pollen contains considerable quantities of fat, but most of it appears to pass through the alimentary canal unaltered, judging by the quantities found among the faeces in the rectum. The very large quantities of fat found in larvae, and the considerable quantities found in adult bees, could be elaborated from absorbed sugar, and bees must possess lipase to enable them to convert and transport fat within their bodies, but that does not necessarily mean that they can digest food fat. Analyses have produced no clear evidence on this point, though contradictory results have been published.

Enormous quantities of pollen and nectar are collected and con-sumed. A colony of moderate size (about 40,000 bees in summer) may collect and use about 100 lb. of pollen annually; it may also make, say, 400 lb. of honey, of which it uses 300 to 350 lb.: the beekeeper's harvest represents the difference between the amounts collected and used, and is thus open to wide variation.

The alimentary canal

The alimentary canal is a tube running from mouth to anus, various regions of its length having specialized structures for different purposes. In the embryos of all animals its origin and early develop-ment is similar up to a point. In the embryos of insects the process goes on as follows, the canal beginning to form in three distinct parts (Fig. 14). The *mesenteron*, or embryonic 'stomach', forms round the remains of the yolk of the egg, in the middle of the body. Small pits appear at the front and rear extremities; these are the orifices of the mouth and anus. The pits deepen, becoming tubular ingrowths, the *stomodeum* and *proctodeum*, respectively. Their inner extremities reach the mesenteron, and eventually join it, forming a continuous tube. The three regions are then the fore-gut, mid-gut and hind-gut. The fore-gut and hind-gut are lined with epithelium which is in fact a continuation of the outer skin of the animal, and it retains the characteristic property of the skin of being able to lay down hard cuticle on its surface. We find this on small bristles, etc., in both the fore- and hind-guts of the adult bee. Further narrow ingrowths from the inner end of the hind-gut become the Malpighian tubules (organs performing the function of kidneys of vertebrates).

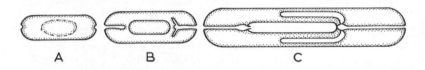

A   B   C

Fig. 14. Diagrams showing the development of the alimentary canal in the embryo. A, the mesenteron (mid-gut) contains the residue of the yolk, and the stomodeum and proctodeum are beginning to appear as depressions at the ends of the embryo. B, the stomodeum and proctodeum continue to sink into the body, and the Malpighian tubules develop from the anterior end of the proctodeum. C, the condition at hatching: the stomodeum (fore-gut) is connected with the mid-gut, and the proctodeum (hind-gut) is in contact with, but not connected to, the mid-gut.

The alimentary canal of the bee larva is in the state shown in the third diagram of Fig. 14. The hind-gut does not connect with the mid-gut until the full-grown larva has taken its last meal and is about to pupate, when the connection is opened, and the excreta accumulated in the mid-gut are voided. The Malpighian tubules are also blocked until this time, when they, too, open and discharge their contents into the hind-gut. Thus is prevented the fouling of the cells and food of the larvae while they are feeding. There is no delay in the junction of the mid- and hind-guts in the larvae of other insects, which, like the caterpillars of butterflies, are active, out-door, creatures. The canal of the bee larva is shown in Plate 19.

In the adult bee the mid-gut is the *ventriculus*, or stomach. The fore-gut becomes differentiated into the mouth cavity, the *oesophagus*, and the *crop*. The hind-gut becomes the *small intestine*, followed by the *rectum*, with the *anus* opening at its end. *Glands*, some of them associated with digestion, deliver their secretions into the mouth.

The mouth

As we have seen in Chapter 2, the opening of the mouth is surrounded by the mouthparts: the labrum and mandibles above, and the proboscis below. Behind the opening, the alimentary canal begins with the *cibarium* or food chamber (Fig. 15). Muscles attached to the inner side of the clypeus and to the walls of the cibarium cause it to dilate when they contract; opposing muscles on the cibarium compress it; the action of these muscles causes the cibarium to act as a sucking pump, which raises fluid through the food canal of the proboscis.

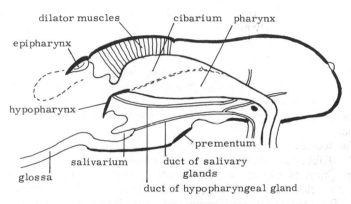

Fig. 15. The cavity of the mouth and the associated structures. A longitudinal section through the head, diagrammatic.

On the floor of the cibarium lies the hardened *hypopharyngeal plate*, the front lobe of which bends downwards. Behind this lobe are the two openings of the ducts of the *hypopharyngeal glands*. From the plate two long slender arms run backwards on either side of the cibarium. They support the cibarium and afford attachments for the muscles by which the cibarium is slung to the frons. Under the hypopharynx is a pouch, the *salivarium*, into which opens the common duct of the *postcerebral* and *thoracic salivary glands*; their secretion runs on to the adjacent root of the glossa, and passes down the glossal canal to mingle with the nectar or syrup which is being taken into the mouth through the food canal of the proboscis. (In the larva the salivarium is the spinneret of the silk glands.)

Behind the cibarium the wide *pharynx* runs towards the vertex, and narrows into the *oesophagus*, which emerges from the head through the foramen. Just before it reaches the foramen, slender muscles sling the canal to the vertex.

The glands

Four sets of paired glands deliver their secretions to the mouth and mouthparts. Though they are not all relevant to feeding and the alimentary canal, it is convenient to consider them together.

The *hypopharyngeal* or *brood-food glands* lie, as their name suggests, above the pharynx and under the frons (Plates 8, 9, 13), and their ducts run under the cibarium to the hypopharyngeal plate.

Each gland consists of a very long, coiled tubule, to which are connected, by short side tubes, several hundred small rounded bodies, the *acini*, the glandular tissue of which produces the fluid brood food. The glands resemble miniature strings of onions (Plate 9). In recently emerged bees, and in those which have over-wintered but have not yet nursed larvae, the acini are plump, cream-coloured organs, and in a dissection present the appearance shown in Plate 13. By the time workers have completed the usual period as nurses, feeding the brood, the acini have shrunk until they have almost disappeared, and they do not again become active. These glands are rudimentary in the queen, and absent from the drone.

The secretion of the hypopharyngeal glands is a sticky, milky fluid. Ribbands (1953) gives a detailed account of the chemical composition of brood food, which he prefers to call 'bee milk', regarding this as a more appropriate name (but it is doubtful if the older, well established name can be superseded). Brood food is fed to all larvae for the first three days, though on the third day honey and pollen are added to the diets of worker and drone larvae. Larvae in queen cells continue to receive brood food only (it is then called 'royal jelly') until they pupate. It is therefore obviously a complete food. It contains several vitamins of the B group, notably acetylcholine and pantothenic acid, these being apparently important for bees. There is currently much analytical research on brood food to some extent initiated by attempts to popularize royal jelly as a panacea—for which, however, there appears to be little if any justification. Brood food runs on to the mandibles, and is placed in cells by them.

The next two sets of glands have been renamed 'labial glands' by Snodgrass, on the grounds that they are derived from ingrowths of H6, the segment to which the labium belongs. The older names used below are, however, descriptive of their position in the body of the adult bee, and are unlikely to be abandoned.

The *postcerebral glands* lie, as their name suggests, behind the brain. The acini are more translucent than the creamy bodies of the brood-food glands, and have a different, characteristic shape (Plate 9). The acini are arranged in small bunches on a branching system of tubules. They run into a median duct under the pharynx, this duct also bringing the secretion of the *thoracic salivary glands*. The latter are rather compact bunches of short cylindrical bodies arranged on branches of the main ducts, and their secretion is stored in two small sacs, from which main ducts pass to the median duct in the head. The site of these glands in the thorax is shown in Plate 9. They are derived from the larval silk glands.

Reported analyses of the products of these two pairs of glands show surprisingly contradictory results, some workers asserting the presence of invertase and others denying it. Similarly the presence of fat, and fat-converting enzyme (lipase) has also been reported and denied. It seems most probable, however, that one (the post-cerebrals) or both of these glands is responsible for producing the enzyme which inverts the sucrose of nectar or syrup, the result of which is the glucose and fructose in honey. The process begins when saliva runs down the glossal canal and is mingled with the fluid which passes up through the food canal of the proboscis. The process of inversion continues, and is completed, in the comb. No further digestion in the bee's alimentary canal is needed, for glucose and fructose can be absorbed through the walls of the intestine, as soon as it is taken in. The only source of fat open to bees is pollen, the grains of which contain large fat globules; judging by the quantity of unaltered fat globules found in the faeces, very little, if any, appears to be assimilated by bees. Bees lay down considerable amounts of fat in their bodies, but this could be made from sugar. Fat is utilized and moved about, however, in the course of normal physiology, and bees must possess the necessary enzymes for fat metabolism.

The fourth pair of glands associated with the mouthparts are the *mandibular glands*. They are single, somewhat lobate, sacs (Plate 13) lying under the genae immediately above the mandibles. A short duct opens in the membrane at the root of the mandible, and the secretion runs into a groove (Plate 3) which leads it to the spoon of the mandible. The glands are of considerable size in the worker, rudimentary in the drone, and very large in the queen. The function of the mandibular glands was a mystery for many years, though the subject of much speculation. Dr. C. G. Butler's discovery of their true significance in the queen was one of the most important and interesting events of recent years. The secretion, called 'queen substance' by Butler, is distributed by grooming, and workers crave for it. The queen's ability to produce it in adequate quantities determines swarming and supersedure by inhibiting queen cell construction, and it seems to prevent the development of ovaries in the workers. Its presence enables workers to recognize their queen and her presence in the nest, and it is a strong factor in consolidating the social unity of the community. When a queen fails to produce queen substance in sufficient quantity, queen cells are built and young queens are raised in preparation for swarming or supersedure, and in the absence of a queen, laying workers appear. Recent work in Brazil suggests that queen substance may also enable drones to recognize a queen on the wing.

The secretion of the glands in workers does not contain queen substance. In their case, the secretion may play a part in the preparation of wax, by kneading in the mandibles, for building comb.

Small *postgenal glands* are situated behind the inner walls of the fossa; their function is unknown.

The alimentary canal in the abdomen

From the head the *oesophagus* passes through the thorax. A long, narrow tube,.it takes up very little room in this part of the body, which is nearly filled by the great indirect wing muscles. The greater part of the canal, with its several specialized regions, lies in the abdomen. It is illustrated in Plates 8 and 9.

In the anterior part of the abdomen, the oesophagus expands into the *crop* ('honey stomach'), a transparent bag which, when full, occupies a large part of the anterior end of the abdomen (Plate 8B). Muscle fibres which surround the crop contract when it is empty and reduce it, by crumpling its thin walls, to small proportions (Plate 9).

The maximum load of nectar that can be carried in the crop is about 100 mg., but the average load brought home is thought to be more like 20 to 40 mg. Thus a pound of nectar represents between 12,000 and 24,000 journeys, and a pound of honey much more.

Also belonging to the fore-gut, though it looks like a projection of the ventriculus into the crop, is the *proventriculus* ('stomach mouth' or 'honey stopper'). This is a valve which prevents the collected nectar from running into the stomach. It also comprises a filtering apparatus for extracting the pollen which is always mixed with nectar, sometimes in large quantities. Four triangular lips

(Fig. 16) on the apex of the proventriculus can be closed or opened. These lips are fringed with fine, closely-set hairs, which are directed backwards, towards the stomach. When nectar is being filtered, the proventriculus makes gulping movements while opening and closing the lips, pollen being caught and retained by the hairs while nectar is returned to the crop. The pollen collects in pouches behind the lips, and when these are full the pollen masses are passed into

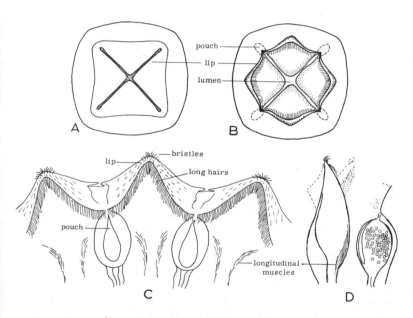

Fig. 16. The proventriculus. A, anterior aspect, lips closed; B, ditto, lips open to show the short spines and long hairs of the lips, and the lumen partly closed by the muscles below the lips, also the pouches which open into the lumen. C, part of the proventriculus laid out after slitting up on one side; three of the four lips are shown, with the pollen pouches between them. D, sketch of a longitudinal section, on the left through a lip, on the right through a pouch.

the stomach. Dr. L. Bailey (1952) has described the method of filtration in detail. The proventriculus can be observed in action in an anaesthetized bee or in a freshly killed one. Behind the lips, muscles close the lips and the lumen of the valve. A short tubular extension of the proventriculus into the ventriculus probably prevents regurgitation of the stomach contents.

The next part of the canal is the *ventriculus* or stomach, derived from the embryonic mesenteron. This is a long, wide tube, lying in a loop, forming a more or less complete spiral turn. The ventriculus is constricted at intervals by contracted circumferential muscles, these constrictions giving it the appearance of the tube of an anti-gas respirator. The muscle fibres, by contracting in sequence from front to rear, produce peristalsis, thus urging the contents of the stomach to move towards the rear end of the organ. The ventriculus is lined with epithelium which is in a state of continuous proliferation, throwing off cells which are glandular in nature (seen in Fig. 17), and which become detached and mingle with the food mass. These cells contain enzymes which digest the proteins in the pollen, reaching them through the germ pores of the grains (for most of the pollen husks found in the faeces are intact and empty, except for fat globules). The epithelium of the ventriculus is the site of attack by the protozoan *Nosema apis*, the cause of 'Nosema disease'.

The next part of the canal is the *small intestine*, this and the rest of the canal being derived from the proctodeum. The small intestine is a narrow tube, surrounded by circumferential muscle fibres. The tube is pleated into six longitudinal folds, and like the ventriculus is coiled. Immediately behind its junction with the ventriculus, the Malpighian tubules open into the intestine. Then follows the *pyloric valve*, formed by the thickening of the walls of the intestine in this region (Fig. 17); this valve regulates the passage of material from the ventriculus into the intestine. The inner surface of the valve bears numerous small bristles, directed towards the rear and probably assisting in urging the contents in this direction. The pleating of the walls of the small intestine greatly increase the area of the surface exposed to passing food, while the reduced area in cross-section must slow down its passage. This is compatible with the view that absorption of digested food takes place in this region of the canal. It does not appear to do so elsewhere, though in the larva it can only be through the wall of the mid-gut.

The small intestine opens into the *rectum*. Like the crop, the rectum is capable of very great distention, also of contraction to a very small pouch. When fully distended, it almost fills the abdomen, then extended and dilated to the limit. In this condition, shown in Plate 8A, it is able to accommodate the large quantity of waste matter which accumulates during long periods of confinement of bees to their nest in winter conditions. Its contracted state is shown in Plate 8B.

The faeces consist very largely of pollen husks, pollen fat globules, and the exhausted epithelial cells of the ventriculus. This matter does not undergo further change in the rectum. If indigestible food has been given to stocks by the beekeeper, who may unwisely give pollen substitutes containing starch at times when bees are confined to their hives by weather, or brown sugar, which contains indigestible constituents, these substances will ferment in the rectum, and bacteria, yeasts, and microfungi will thrive in enormous numbers. The resulting irritation has two bad effects: it causes bees to fidget, thus raising the temperature—sometimes sufficiently to bring on premature egg laying—and it may cause dysentery.

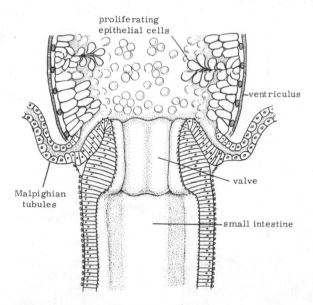

Fig. 17.   The pyloric region.   Half of the canal is cut away, showing the interior of the other half, with the junction of the ventriculus and the small intestine, and the valve, also the insertion of the Malpighian tubules. Note the small recurved setae lining the valve, also the proliferating digestive cells of the ventricular epithelium.

Six *rectal pads* (not 'papillae', an unsuitable term), partly chitinized, are arranged round the rectum and in its walls.   They appear to be vestigial, with no function.   It has been suggested that they extract water from the contents of the rectum and return it to the body, but their microscopic structure does not support this view.   Other insects which live in arid conditions must conserve their body moisture, and in some of these the rectal pads (or 'rectal glands') may have this function, though it is certainly performed by specialized regions of the Malpighian tubules in many insects. Bees are not normally short of water, and probably need no means of conserving it.

It has also been suggested that the rectal pads absorb and pass

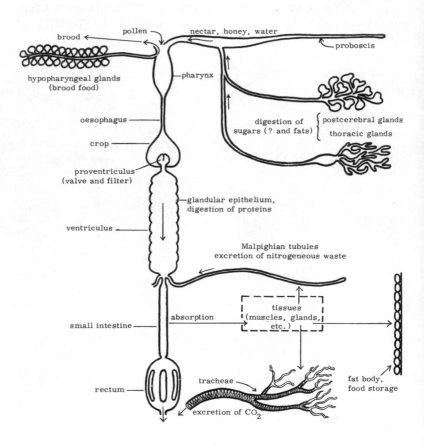

Fig. 18. Diagram of the alimentary canal, with relevant glands, etc., and the physiology of nutrition so far as it is known.

into the blood digested fats, and that they may maintain at a steady level the concentration of salt in the blood, but these views need confirmation.

The rectum tapers to the anus, an aperture in the proctiger.

The Malpighian tubules (Plate 9, Figs. 14, 17)

The Malpighian tubules are offshoots of the proctodeum, and are the principal organs of excretion. Their appearance in the embryo as ingrowths from the proctodeum indicates clearly that they belong to the hind-gut, and their replacement in the pupa by the adult tubules confirms this fact, which refutes the theory, advanced by some, that they belong to the mid-gut. They number about one hundred, and are long slender tubes, closed at their distal ends, their walls composed of a single layer of cells. Through their walls nitrogenous waste is absorbed from the blood and passes down the tubules and into the small intestine. The tubules are the site of attack by the protozoan parasite *Malpighamoeba mellificae*, the cause of 'Amoeba disease'.

The fat body

The storage of food in the body of the bee is an interesting aspect of nutrition. The *fat body* is a layer of conspicuous creamy cells (Plate 10) concentrated principally on the floor and roof of the abdomen. There are two kinds of cell, mixed together. The *fat cells* contain much fat, and also at times protein (albumen), the latter being quickly converted into brood food, and not accumulating except at times when brood is not being fed. Glycogen ('animal starch'), the only kind of carbohydrate which animals can manufacture, is also stored by bees in their fat cells, where it serves as reserve energy food: it can be converted into glucose in a short time, *e.g.*, while a travelling bee rests, and is then ready for use at once.

The second kind of cells in the fat body are *oenocytes*, which are thought to have some function connected with wax production, perhaps as producers of some necessary enzyme. The fat body is particularly well developed over the wax glands.

The diagram, Fig. 18, summarizes the physiology of nutrition so far as it is known.

CHAPTER 4

THE BLOOD AND CIRCULATION

IN insects the blood carries no solid red corpuscles, the haemoglobin of which collects oxygen and liberates waste carbon dioxide at the lungs of vertebrates. Insects have no lungs, and only a very few have haemoglobin or any similar substance in solution in their blood. The bee has none, and her blood has only a localized respiratory function, as we shall see in the next chapter. Its main function is to transport food material which has been absorbed from

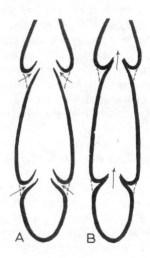

Fig. 19. Diagram illustrating the action of the heart, and showing the blind posterior chamber, and the one in front of it. A, the heart dilated, ostia open, blood entering. B, the heart contracted, ostia closed and acting as valves, blood driven forwards.

the intestine to the tissues that need it, and to carry away waste, also in solution, to the organs of excretion. The almost colourless plasma contains phagocytes, white corpuscles which ingest and destroy invading bacteria.

Blood fills the body cavity of the bee; it is not confined in a closed system of arteries, veins, and capillaries, like ours, but there is a pulsating heart, and also other organs which assist circulation; and there is one large blood vessel and one or two small ones.

In the adult bee the *heart* (Plate 8) is an elongated organ lying just under the roof of the abdomen and attached also to the dorsal diaphragm (see below). It has muscular walls, and is pierced by five pairs of openings, the *ostia*, with one-way valves which allow blood to enter the heart when it is dilated, but confine it, and force it forward, when the heart contracts, as shown diagrammatically in Fig. 19.

The *aorta* is continuous with the heart. It is a blood vessel which runs through the thorax to the head, where its end opens below the brain, which is thus provided with a steady supply of food. Just after entering the thorax, the aorta is repeatedly looped, but it is a simple narrow tube when it passes between the indirect flight muscles. There is another small pulsating vesicle under the bases

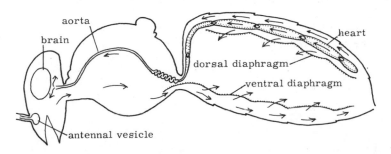

Fig. 20. Diagram illustrating the action of the heart and diaphragms.

of the antennae; from it two small vessels run into the antennae as far as their tips.

The *dorsal and ventral diaphragms* (Fig. 20) are responsible for setting up a circulation inside the abdomen and for drawing blood from the thorax into the abdomen. These diaphragms are sheets of very thin, transparent membrane which spread over the floor and roof of the abdomen, and are attached to the apodemes of the tergites and sternites, but between these points the edges of the diaphragms are free. Radiating from the points of attachment are muscle fibres. A large bed-sheet, suspended from the ceiling by drawing pins fixed at intervals of two or three feet, and sagging between the pins, would make a rough model of the dorsal diaphragm.

The dorsal diaphragm extends from A3 to A6, the ventral diaphragm from T2+3 to A7, while two narrow strips extend as far as the spiracle plate of A8. Both diaphragms pulsate by a sequence of contractions of their muscles, which sets up a series of waves in the sheets of membrane; these waves create currents in the blood which lies between them and the body wall to which they are attached (these spaces are called the dorsal and ventral *sinuses*). The ventral pulsations drive the blood in the ventral sinus from the thorax into the abdomen, and on towards the rear. The driven blood escapes from the sinus between the points of attachment of the diaphragm. The dorsal diaphragm works in the opposite direction, driving the blood in the dorsal sinus forwards, towards the thorax, and it, too, escapes from the sinus between the points where the membrane is attached to the apodemes. Thus the blood in the abdomen is kept circulating forwards in the dorsal part and backwards in the ventral part, while some of the blood, escaping from the loose edges of the diaphragms, sets up eddies in all directions. Blood drawn from the thorax is replaced by blood forced into the head by the aorta. Blood enters the legs, the cavities of which are continuous with the general body cavity, and is kept in circulation by the leg movements.

CHAPTER 5

THE RESPIRATORY SYSTEM

INSECTS have no centralized respiratory organs corresponding with
our lungs. Air is brought from the outer atmosphere direct to the
tissues, entering the body through apertures in the body wall and
passing through a system of air pumps and ramifying tubes. The

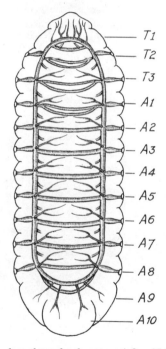

Fig. 21. The larval tracheal system (after Nelson).

whole apparatus is something like the forced draught ventilating
system of a large ship, though it has no inner vents.

The plan of the adult bee's respiratory system is clearly revealed
by the larval equipment (Fig. 21), which consists of longitudinal
main trunks running along each side of the body, with short lateral
branches connecting them with the segmental *spiracles* (breathing

57

holes), and connected with one another by ventral transverse *commissures* in each segment. These tubes are called *tracheae* because, in both larva and imago, they are maintained in a dilated state by spiral thickenings of cuticle (*taenidia*) in their walls, and thus resemble the windpipe (trachea) of terrestrial vertebrates, which has similar thickenings serving the same purpose.

In the adult the same general plan is retained, with considerable modifications in the thorax and head. The longitudinal trunks are

Fig. 22.   The principal tracheal sacs and trunks of the adult bee.

expanded into large air bags, *tracheal sacs* (Fig. 22), especially in the abdomen, where they are very large in A2 and A3, but taper off towards the rear (Fig. 22). The transverse commissures are also inflated, as well as some main branches. Similar sacs occur in the rear part of the thorax, and a sac surrounds the brain. Large unmodified trunks come from the first spiracle and send main branches to the head and thorax.

The tracheal sacs serve as bellows, contracting under the pressure of the surrounding blood when the abdomen is retracted and compressed, and expanding when it is extended and dilated. The rapid, rhythmic pulsations of the bee's abdomen are thus respiratory movements.

From the sacs and the large trunks smaller branches ramify to every part of the body and of each organ. The final, very small branches have no spiral thickening and are called *tracheoles*. A network of tracheoles is closely applied to the tissues; they are open at their extremities, and there they contain a little of the surrounding blood plasma. Oxygen brought in to the tracheoles dissolves in the blood, and is thus at once available to the tissues in the necessary form, *i.e.*, in solution; and carbon dioxide, the result of the oxidation processes of metabolism, is removed by the same means. The ultimate method of gas exchange is thus the same, in principle, as our own, the difference between the two systems lying in the relative importance of the blood and the ventilating trunks in transporting the gases. The tracheal method is suitable only for small animals. It could not serve a large body efficiently.

In both the larva and the imago there are ten pairs of spiracles, one pair per segment from T2 to A8. All except that of T3, which is very small, have valves which are operated by muscles. The valve of the first spiracle (T2) cannot be completely closed, which probably explains why the acarine mites are able to enter this spiracle but none of the others. The propodeal spiracle (A1) is the largest. Like the other abdominal spiracles, its valve has both opening and closing muscles, and is situated at the inner end of a short *atrium* or porch.

When the bee is in flight, she requires very abundant supplies of oxygen, and must be able to get rid of similarly large quantities of carbon dioxide. Dr. L. Bailey (1954) has shown that, at such times of maximum activity, air is inhaled at the first spiracle (T2) and exhaled mainly at the propodeal spiracle, the adjacent great indirect flight muscles being thus catered for. Some air is inhaled at the abdominal spiracles also. The need for free intake of air at the first spiracle during flight explains why bees suffering from acarine disease, when the infesting mites block the trachea, are unable to fly. When bees are inactive it appears that air is both drawn in and driven out mainly at the first spiracle.

Though the bee's blood contains nothing like haemoglobin, the indirect wing muscles contain *cytochrome*, which facilitates gas exchange; this substance gives these muscles their pink colour.

Insects can tolerate much higher concentrations of carbon dioxide than we can. The gas is used as an anaesthetic for queen bees during instrumental insemination; its only effect is to accelerate maturing, the queen beginning to lay in a very short time. Virgin queens treated with carbon dioxide three to four days after emerging from their cells begin to lay drone eggs without delay, this fact being utilized, in genetical experiments, to make it possible to fertilize a queen with spermatozoa from her own sons, a short cut to the production of a homozygous strain. Young workers treated with the gas cease their domestic duties and begin foraging.

CHAPTER 6

CONTROL: THE NERVOUS SYSTEM AND SENSE ORGANS

THE bee has what we call a 'brain', but it barely resembles ours. Our brain, enormously developed, is the seat of consciousness and intelligence. Another part of it has automatic functions of which we are quite unconscious. Imagine all of our brain removed except a microscopic bunch of cells, and with all its automatic functions transferred to the spinal cord, which in turn is reduced to barely visible proportions, and you will have a model of the bee's brain.

There is no evidence of real consciousness and intelligence in bees. The physical apparatus for it does not exist. There is, however, a small but admirable mechanism of nervous co-ordination in the working of tropisms, or response to external stimuli received by the sense organs. All the bee's actions, though they may at first sight suggest intelligence and reasoning, are in fact purely automatic responses to external conditions and to internal chemical conditions. If the bee could think and reason, she would not be so easily deceived as she is by the beekeepers who interfere with her in their systems of management; they create the conditions which they know will produce automatic results; they do not always succeed, because as yet they do not know enough about the correct stimuli to apply. Bees have limited and short-lived powers of memory.

The nervous system

In the embryo, the primitive beginnings of the nervous system are derived, as in other animals, from infoldings of the ectoderm, the outer layer of cells. The primitive system consisted of a knot of nervous tissue or *ganglion* in each segment, with radiating nerve fibres running to the muscles and other segmental organs. Thus each segment was almost completely autonomous. This plan is still clearly shown by the larval outfit (Fig. 23), though considerable modification has already occurred. The ganglia of the first six segments are already combined to form the lobes of the brain and the suboesophageal ganglion, and those of A8 to A10 are fused together, and now constitute a single ganglion situated in A8, but sending nerves to A9 and A10 also. The ganglia are connected together by twin nerve trunks, the *longitudinal commissures*.

Further condensation of the ganglia occurs during pupation. The adult system is indicated in Fig. 23, and is shown in Plate 10. The ganglia of T2, T3, A1, and A2 are now combined to form the large second ganglion of the thorax, and that of A7 has now joined the already compound ganglion of A8-10. The central nervous system of the adult thus consists of the brain (H1 to H3), sending nerves to the simple and compound eyes and antennae, also to the labrum

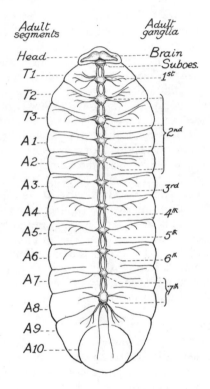

Fig. 23. The larval nervous system. The six ganglia of the head are already combined to form the brain and suboesophageal ganglion. Those of T2 and T3, and A1 and A2, are still discrete; that of A7 is still discrete, but those of A8 to A10 are already fused (after Nelson).

and cibarium, the *suboesophageal ganglion* (H4 to H6), close to the brain but below the oesophagus, sending nerves to the mandibles and proboscis; the first thoracic ganglion (T2), nerves from which go to the forelegs; the second thoracic ganglion (T2 to A2), controlling all these segments, its nerves going to all the wing muscles, and the second and third pair of legs, some passing through the petiole into A2; in the abdomen are five more ganglia, the first four belonging to A3 to A6, and the last to A7 to A10. The last is firmly fixed to the sting apparatus, and, as we have seen, always comes away with it when it is torn out of the body. In the drone and queen this ganglion controls the movements of the reproductive organs, and in the queen the process of egg laying.

The segmental ganglia still retain independent control of their

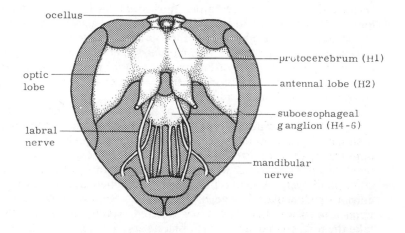

Fig. 24. The brain and principal nerves of the head, anterior aspect. The labral nerves come from the very small tritocerebrum (H3), concealed behind the antennal lobes. The mandibular nerve comes from H4. Two other pairs of nerves (shown but not flagged) come from H5 and H6, and go to the maxillae and labium respectively.

organs to a considerable degree. An insect which has been beheaded can move its legs and wings vigorously, but it is no longer able to co-ordinate its movements. The abdomen of a female moth, after the petiole has been severed, can still go on laying eggs.

The *brain*, and the principal nerves coming from it, are shown in Fig. 24, and it is also seen in relation to other organs in Plates 8 to

10 and 13. Its three component parts, belonging to the first three segments of the head, can be distinguished; they are called, respectively, the *protocerebrum*, *deutocerebrum*, and *tritocerebrum*. The greater part of the protocerebrum is made up of the very large optic lobes, great bundles of crossing and recrossing nerve fibres (*chiasmata*) connected with the thousands of units of the compound eyes. The deutocerebrum is similarly made up of bundles of nerve fibres connected with the sense organs of the antennae. The tritocerebrum is small and obscured by the other parts; it sends nerves to the labrum and frons (for the cibarium).

Buried in the protocerebrum, under the ocelli, are the *corpora pedunculata* ('stalked bodies', also called the 'mushroom bodies' because they somewhat resemble mushrooms in form). They contain small groups of nerve cells and are the *association centres*, connected by nerve fibres with the optic and antennal lobes and also with the other parts of the nervous system. They perform the very important rôle of co-ordinating the actions of the insect according to the information received from the sense organs—the 'intelligence service' of the animal.

As we might expect, on account of her many skills and activities, this part of the brain is larger in the worker bee than in the other castes, and is smallest in the drone. The drone's brain has a deceptive appearance of large size, but this is due to the extensive development of his optic lobes, which are in proportion to his very large compound eyes. A method of estimating ability by measuring the corpora pedunculata has been formulated. Results obtained by its application show that queens of different species of bumble bees take the first, second, and fourth places, and a queen wasp the third, the honeybee worker coming fifth. A method of this kind is always suspect, and likely to produce very misleading results, but these are not altogether surprising. The bumble bee and wasp queens have to perform all kinds of work, whereas in the honeybee community there is differentiation of function in the castes.

The sense organs

The organs of the bee which receive and transmit to her brain information about her environment are quite different from our own. From observation and simple experiments we know that the bee can see, feel, taste and smell; the last two senses are identical over a large range, as in our own bodies. There is no clear evidence that the bee can hear. The most important sense organs are obviously the compound eyes and the antennae, but the detection of stimuli is by no means restricted to these organs, for the units of sensual perception are also found, in large numbers, all over the bodies of insects. It is, perhaps, a little difficult for us to get used to the idea that some insects taste with their feet!

The eyes (Plate 2)

The *ocelli*, or simple eyes, are not highly organized. An ocellus (Fig. 25) consists of a lens above a layer of very simple, elongated retinal cells connected to nerve fibres. The retina has none of the elaborations of the vertebrate retina. No image can be formed on the bee's retina, and it is assumed that its only function can be to detect the relative intensity of the light which falls on the lens.

The *compound eyes*, however, are very complicated structures.

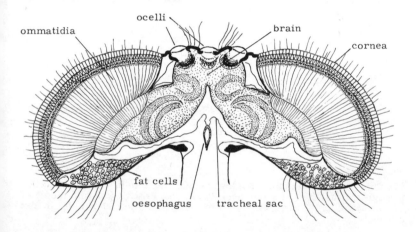

Fig. 25.   Transverse section of the head of a drone. Drawn from a microtome section.

The external surface of each eye is an elongated oval, strongly convex, and it consists of the lenses of thousands of units, the *ommatidia* (Fig. 25). The number of units in one eye is probably variable. According to the writer's measurements and estimates, performed on a small number of individuals, the worker's eye has an area of about 2·6 sq. mm., with about 6,900 lenses, each about 0·2 mm. diameter. The queen has fewer, the drone a great many more, though not so many as might be expected, for the lenses themselves are broader. Each drone cornea has about 8,600 lenses, diameter 0·3 mm., in an area of about 9 sq. mm.

The ommatidia (Fig. 26) are elongated bodies, tapering towards their inner ends. They radiate in directions perpendicular to the surface of the cornea, and thus each one covers a very small angular field of view. An ommatidium consists of a *lens*, behind which is a clear, transparent *crystalline cone*, surrounded by pigment cells. Behind the cone, and in contact with its apex, is a bundle of eight long *retinula* cells, also surrounded by pigment cells. The pigment cells of the ommatidium appear to serve to exclude the light which enters neighbouring ommatidia, thus ensuring that stimulation is applied only by the light entering the individual unit. The edges of the retinula cells, which meet in the axis of the ommatidium, combine to form a long narrow *rhabdom*, like a transparent rod running back to the proximal end. The rhabdom is striated, and it appears that its function is to divert the light which travels through

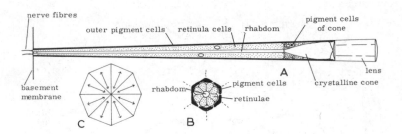

Fig. 26. An ommatidium. A, longitudinal section; B, transverse section through the rhabdom and retinula cells; C, von Frisch's polaroid model of B, consisting of 8 triangles with axes of polarization shown by double arrows.

it laterally into the cells. At the narrow, proximal end of the rhabdom, nerve fibres from each of the eight retinula cells pass through the basement membrane and into the optic lobes of the brain, where they cross and recross in complex chiasmata.

The compound eyes cannot form images, such as are produced by the lens of a vertebrate eye on its retina. It is assumed, therefore, that each ommatidium detects the intensity of light in the field immediately in front of its lens, and that the total impression received by the complete eye is in the nature of a mosaic of small dots of varying degrees of brightness, very similar to the mosaics printed by coarse half-tone blocks in newspaper illustrations. Though capable of only limited sharpness of definition (estimated at about 1/100th of the acuity of human sight), the compound eye is efficient enough to enable the bee to recognize landmarks as she approaches the hive. It is also well adapted to detecting movement, which changes the mosaic pattern. The bee's 'fusion frequency', a measure of sensitivity to flicker, is about 300 per second, so she can recognize the shape of objects while she is in rapid flight. (Human fusion frequency is only about 30 per second; this determines the minimum speed of running cine film through the projector.)

It has been repeatedly shown by experiment that bees can perceive differences in colour of the objects which they see; but their colour perception is not the same as ours. Fig. 27 shows the parts of the spectrum visible to the human and the bee's eyes. The bee is blind to the red end of our spectrum, but can see by ultra-violet, which

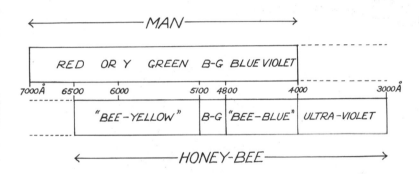

Fig. 27. The visible spectra of man and bee. OR, orange; Y, yellow; B-G, blue-green. Wavelengths are shown in Angstrom units.

we cannot. The bee may not have acute powers of distinguishing between colours, but can see four broad bands, 'bee yellow', blue-green, 'bee blue' and ultra-violet. Though blind to red, bees react strongly to some red flowers because these reflect ultra-violet rays, which are attractive to bees. Red flowers without UV reflections appear as black to bees. White lead paint reflects UV, and therefore gives the bee light over the whole range of her perception ('bee white'). Zinc white does not reflect UV, and must give the bee a blue-green sensation. White narcissi reflect light as white lead does and produce a similar stimulus.

The compound eyes have been proved to possess another property lacking in our eyes: they can detect the plane of vibration of polarized light. (A brief explanation of this phenomenon is included in

Appendix I.) The light coming from clear sky is partly polarized, and the plane of vibration varies according to the angle between the direction of the sun and the direction of the particular quarter of the sky in question. Von Frisch and his colleagues have shown that bees find their way home, and register the direction in which a new nectar source lies, by reference to the direction of the sun, or if the sun is not visible, by the plane of vibration of light; also that scouts who discover new nectar sources can communicate their direction by means of the dances which they perform on the

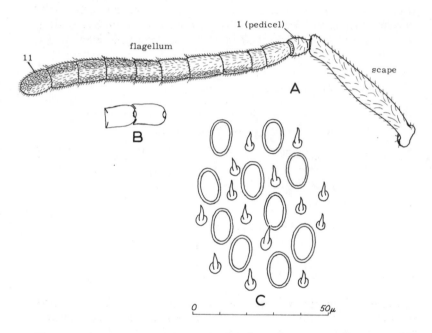

Fig. 28. A, antenna of drone; B, jointing of segments; C, sense plates and sense hairs on the antenna, as seen with a 4 mm. objective.

combs. It is believed that this power of detecting the plane of vibration resides in the retinula cells. A section through an ommatidium is shown in Fig. 26, and by its side a diagram of von Frisch's 'artificial eye', a model of this cross-section (see Appendix I), made of eight triangles of polaroid representing the eight retinula cells, and arranged with their axes of polarization as shown by double-ended arrows. When this model is held up before the eyes towards a patch of blue sky, a pattern of varying light intensity will be seen in the triangles. If, in fact, the retinula cells are sensitive in this way, their messages to the optic lobes could be a navigational aid; they would act as a built-in sun compass. Ants also possess this faculty; so do crustacea: the primitive king crab has been useful in experimental work because its ommatidia are very large, and it has been possible to record differences in electrical reaction of the retinula cells.

The antennae (Plate 2)

The antenna of the bee (Fig. 28) consists of the *scape* and the *flagellum*, the latter having 11 segments in the worker and queen

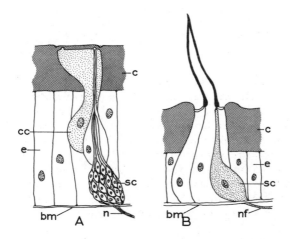

Fig. 29. Sensilla. A, a sense plate in section; B, a sense hair. *c*, cuticle; *cc*, cap cell; *sc*, sense cells; *e*, epidermis; *bm*, basement membrane; *n*, nerve; *nf*, nerve fibre.

and 12 in the drone; the antenna of the drone is also longer and broader than the worker's. Several different kinds of sense organs, or *sensilla*, are borne on the antenna, and some of these are present in very large numbers, particularly towards the tip of the flagellum. On a single drone antenna there are 500,000 sensilla, of which about 30,000 are sense plates. A worker's antenna has 5,000 to 6,000 sense plates, and a queen's only 2,000 to 3,000. Some sensilla also occur in the mouth and elsewhere on the body, often in large numbers.

A *sensillum* is composed of a sense cell or cells with a nerve

E

fibre connected with the central nervous system, and its other, distal end in close connection with the cuticle; associated with it is a structure derived from the cuticle, and one or two accessory cells. Sometimes there is a group of sensory cells. This is the general plan, but in detail there are several types of sensillum. *Sense hairs* are sensilla with a fine bristle projecting from the cuticle; *sense pegs* are similar, but have a short, stout peg instead of the bristle. These are assumed to be tactile organs. They occur in large numbers all over the antenna, and the pegs are common near the tip of it. These and other types are illustrated in Fig. 29. *Pit hairs* and *pit pegs* are similar structures buried in pits, with their tips just below the narrow mouths of the pits. They are said to be organs of taste and smell, but there is no direct evidence of that. Minute *sense pores*, in which the tips of sensory cells are exposed, are scattered in great numbers all over the surface of the body, but occur only on the pedicel on the antenna. *Sense plates* are found in great numbers on the segments of the final eight segments of the flagellum, thickly on the terminal segments. They are lacking from a strip along the surface of the antenna which faces outwards. These sensilla consist of a hollow in the cuticle, capped by a thin plate level with the surrounding surface, and like the head of a drum; this is in close connection with a slender process of the sensory cells. These are obviously important detectors, but their function is not clear. They have been said to .be organs of taste and smell, and of hearing, but they seem to be unsuitable for all these functions.

The antennae have been shown, beyond all doubt, to be the principal, though not the only seats of the senses of taste and smell, so some, if not all of the sensilla found on them must be the relevant detectors.

The '*organ of Johnston*' is situated in the pedicel. It consists of a group of sensory cells arranged round the nerve trunk which runs through the antenna, and the tips of the sensory cells reach the intersegmental membrane round the joint above the pedicel. It is believed to detect and analyse vibrations of the antenna, and thus to be a speed-of-flight indicator.

Scolophores are another kind of sensilla which occur in the legs and head of the bee. They consist of a chain of three cells, of which the sensory cell is the lowest; it sends a fine process through the two cells above it, as far as the cuticle. These sensilla are supposed to record stresses and strains in the exoskeleton. They were once thought to be auditory organs, but there is nothing to suggest such a function.

Knowledge of the functions of the sensilla is very incomplete, but this is not surprising, for the extremely small size of the organs makes experimentation very difficult indeed.

Though there are no recognizable organs of hearing in the bee, it seems likely that the whole of the exoskeleton might well serve as a tympanum, sensitive to vibrations, and communicating them to nerves ending in the basement membrane. As to the noises made by bees, buzzing has been proved to be caused by vibration of the wings, humming by vibration of the spiracle valves. Bees at the hive entrance, and at nectar sources, emit supersonic noises near the threshold of sound audible to man. The production of piping sounds by queens has been studied by E. F. Woods (1956, 1959). At the time of going to press, an important book on insect sounds, by P. T. Haskell (1961), has appeared. There is very little evidence of any reaction by bees to sounds, though they are known to dislike the proximity of lawn mowers.

Individual bees have no direct means of controlling their body temperatures. They quickly assume a temperature near to that of the ambient air, and their activity, both physical and physiological, quickens as temperature rises and slows as it falls. This automatic effect of temperature on metabolism may serve to stimulate equally automatic social actions to adjust temperature, like clustering to conserve heat in the winter, sealing crevices with propolis, and utilizing the cooling effects of evaporation of water to reduce the temperature of the hive in summer. No special structures like sensilla would be necessary. Bees appear to be able to detect changes in humidity, again, perhaps, by similar mechanisms.

The many experiments which have been conducted seem to indicate that bees have similar tastes to our own within their restricted diet, though with some exceptions. They are averse to peppermint, and like potassium ferrocyanide. Their favourite food is honey, and they can distinguish between different kinds.

CHAPTER 7

THE REPRODUCTIVE ORGANS

Reproductive organs of the drone (Plates 14, 15)

In most other insects the external organs ('claspers') have some function in copulation, but in the honeybees (the genus *Apis*) they are reduced to small sclerites attached to the sternite of A9 (Plate 15); they are homologous with some of the sting parts of the female, *i.e.*, those derived from A9. On the other hand, the internal intromittent organ or endophallus (performing the function of a penis) is very large, and does not resemble that of other Hymenoptera.

The *testes* are composed of bundles of tubules in which the spermatozoa are produced and matured. In drones which have just emerged from their cells the testes are enormous, white, bean-shaped bodies which occupy a large part of the space in the abdomen (Plate 14), but by the time the drone is mature (about 13 days after emergence) they are reduced to small, greenish-yellow scraps of tissue (Plate 15), all their contents having passed on through the coiled tubes of the *vasa deferentia* into the *seminal vesicles*. These are two curved, sausage-shaped organs which increase in length and girth as they receive spermatozoa from the testes (compare Plates 14 and 15). Their muscular walls are lined with glandular tissue, which breaks down to form the seminal secretion; but spiral projecting ledges remain on the inner surface of the walls, and spermatozoa cling to these by their heads. Finally, the vesicles are packed with spermatozoa and the lymph-like secretion.

The accessory or *mucus glands* are very large club-shaped sacs filled with mucus, joined at their bases to make a large U (Plate 15), where the seminal vesicles are connected by short narrow tubes. Like the seminal vesicles, the mucus glands increase in size as the drone matures.

The *ejaculatory duct*, through which the spermatozoa pass at the moment of copulation, springs from the middle of the U near to the openings of the seminal vesicles. These openings are brought close to the opening of the duct when the seminal vesicles are emptied, and the semen passes into the duct, followed by the contents of the mucus glands.

The spermatozoa are remarkable objects (Fig. 30). They have the form of very slender threads, about ¼ mm. long and $0·5\mu$

(1/2,000 mm.) in diameter. Of this length, about 10μ is differentiated into a head, containing the nucleus, which carries all the genetical characteristics transmitted by the drone. The posterior 4·5μ of the head, and the axial part only of the tail and the front half of the head, take up stains readily, which accounts for the appearance of the sperms in Fig. 30A. Spermatozoa swim by lashing their tails. In spite of its extremely small diameter, the tail contains 9 fibrils

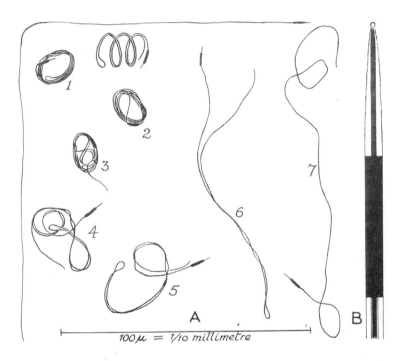

Fig. 30. A, spermatozoa, as they appear in a stained smear: 1, 2, coiled, inactive; 3 to 7, stages in uncoiling. The total length of a spermatozoon is about ¼ mm.; the head is about 10μ long and 0·5μ in diameter. B, structure of the head and part of the tail. (A, drawn from smear; B, simplified after Rothschild.)

and 2 mitochondrial strands (observed by Lord Rothschild with the electron microscope). In the seminal vesicles and spermatheca the sperms lie in swathes, side by side, but when extracted and transferred to water on a microscope slide, they coil up in a systematic manner, the tail being doubled and then thrown into 3 coils, forming a tight hank. They also assume this coiled form when they are naturally ejaculated, but uncoil and rearrange themselves in swathes, with their tails parallel, when they come to rest in the queen's oviducts (see below, p. 78). They are robust organisms, resisting desiccation and living for a long time when frozen in normal salt solution.

The rest of the drone's genital apparatus, including the ejaculatory duct, constitutes the endophallus. The long duct joins the *bulb* of the endophallus, an ovoid body with crescent-shaped and roughly triangular sclerotized plates in its walls and trough-like internal projections. A somewhat contorted tube, the *cervix*, follows, decorated with a double, fringed lobe and with an arrangement of plates on both dorsal and ventral surfaces, the significance of which is not known, if they have any at all. Finally conspicuous horns (*pneumophyses* or *cornua*) spring from each side of the *vestibule*, the final part of the tube, which ends at the *phallotreme*, or genital opening in A9.

At the moment of copulation, a violent contraction and compression of the drone's abdomen, produced by the very powerful abdominal muscles, causes the endophallus to be everted, *i.e.*, turned inside out, just as we blow out the fingers of a glove which, by the action of removing the glove, have been turned outside in; a closer and familiar model is the child's toy which consists of a rubber ball painted like a face, with a mouth from which a long red tongue is forced when the ball is squeezed. Almost simultaneously with eversion, the seminal vesicles contract, forcing semen into the ejaculatory duct and bulb, and it is followed by the mucus from the mucus glands. As eversion takes place, the bulb passes down the cervix, followed by the duct. Though it has been stated that the bulb itself is turned inside out, other observers, including Snodgrass, have found uneverted bulbs in the bodies of fertilized queens after mating; and there would seem to be no special reason why the bulb should be everted. There is a great deal of doubt about precisely what happens. We shall resume consideration of this sequence of events later, in the section on mating.

The slightest stimulus will cause eversion. If a mature drone is handled, the slightest pressure will cause the abdomen to contract violently, whereupon the endophallus shoots out with an audible 'pop'. Killing a mature drone, by any means whatever, always produces eversion, so it is not possible to dissect a fully mature drone with his endophallus undisturbed. Partly matured drones, in the condition shown in Plate 15, often show the ejaculatory duct swollen with semen and mucus which have already passed into it from the seminal vesicles and mucus glands. It is not known whether this occurs normally in maturing drones, in preparation for mating, or whether it occurs when the drone is killed, without eversion taking place. Eversion resulting from death usually reaches only its first stage (Plate 15), but when it results from handling, the further stage shown in the Plate can be brought about by continued firm pressure on the sides of the abdomen. In natural mating, perhaps this completed eversion is brought about by the violence of the struggle between the queen and the drone.

Reproductive organs of the queen (Plate 16)

The *ovaries* of the queen are paired organs, each one consisting of a bundle of about 150 or more tubules, the *ovarioles*. The anterior ends of the ovarioles are thin threads, adhering together and attached to the ventral side of the heart (but shown detached in the Plate),

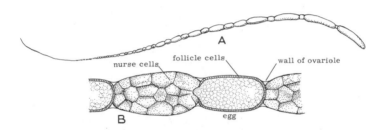

Fig. 31. A, an ovariole with eggs and nurse cells, in all stages of development. B, an egg with its nurse cells, drawn from a microtome section; a plug of the egg's cell plasm is in direct contact with the nurse cells through an opening in the layer of follicle cells.

at the anterior end of the abdomen. In the fertile queen, the ovarioles increase in diameter behind their tips, and the ovaries are very large, filling a large part of the abdomen. The Plate shows them both undisturbed and laid out to reveal the remainder of the apparatus. Egg cells are budded off from the germinal tissue in the tips of the tubules, and begin to slide down the tubules. As they go, they become differentiated into three kinds of cell: the true *egg cells*, *nurse cells*, and small *follicle cells* which cluster round the egg cells and form a continuous sheath round each, except at the forward ends. Now egg cells alternate with masses of nurse cells, the protoplasm of each egg being in direct contact with the nurse cells which follow it. As they continue to pass down the ovariole, the nurse cells absorb food through the transparent, delicate walls of the ovariole, and increase in size, while the egg takes nutrients from the nurse cells and also quickly grows larger. A half-grown egg, enclosed in its follicle and in contact with its nurse cells, is shown in Fig. 31. Growth continues to a point when the nurse cells begin to shrink, and eventually, as the egg approaches full size, the last of the nurse cells is absorbed. The follicle cells also disappear, leaving a fine network of markings on the exterior of the egg. At the apex of the egg, where it was not covered by follicle cells, a small area, the *micropyle*, remains naked, or covered only by an exceedingly thin membrane. It is through the micropyle that spermatozoa penetrate when the egg is fertilized.

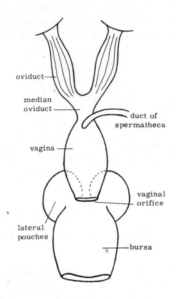

oviduct

median
oviduct

duct of
spermatheca

vagina

vaginal
orifice

lateral
pouches

bursa

Fig. 32. The reproductive tract of a queen, dorsal aspect. Diagrammatic.

At their posterior ends the ovarioles open into the *lateral oviducts* (though their ends are blocked by membrane until the queen begins to lay). The broad, rather short oviducts meet when they join the *median oviduct* (Plate 16 and Fig. 32). Eggs pass from the ovarioles into the oviducts, following one another closely. The median oviduct opens into the *vagina*. The opening of the posterior end of the vagina is in the form of a horizontal slit, beyond which lies the *bursa*, a wide membranous pouch at the anterior end of the sting chamber; it is firmly attached above to the sting and below to the floor of the abdomen. It originates as an ingrowth from the sternite of A8, or of the intersegmental membrane between A8 and A9. At each side of the bursa is a lateral pouch, the significance of which is not known. It used to be thought that the cornua of the endophallus entered these pouches during mating, but apparently this is not so.

Mating

The queen and drone mate on the wing. Many attempts have been made to control mating, and thus to perpetuate good strains or to obtain desirable new ones, by liberating selected drones and queens in large glasshouses, but without success. The reason for these failures has recently been discovered by N. E. Gary, of Cornell University, whose paper describing his investigations will appear in *Bee World* before this book is published. He found that drones and queens will not couple at heights above the ground of less than 35 or 40 feet, but that above that altitude mating occurs freely, even when queens are *tethered*. This important fact may well have far-reaching results in the development of new techniques of controlled breeding.

There has been much controversy and speculation concerning the process of copulation, which is very difficult to observe in natural conditions. Dr. Gary's work puts an end to all doubts. The drone mounts on the queen's back and his endophallus is inserted into the queen's sting chamber in the first stage of eversion. The drone then falls backwards in a paralysed condition, and as he falls the second stage of eversion is reached with explosive violence, after which he drops to the ground, leaving his severed endophallus in the queen's sting chamber.

The detached endophallus, partly protruding from the tip of the queen's abdomen, is the 'mating sign'—the queen's marriage certificate, as Snodgrass calls it!—which can be seen when the queen returns to the alighting board of her hive.

The end of the endophallus probably reaches as far as the bursa, and semen is forced into the vagina and thence into the oviducts, which may be completely filled with it; part of the endophallus remains in the sting chamber, embedded in mucus. The drone's mucus has the property of coagulating immediately it comes into contact with air; thus, with the bulb of the endophallus, it forms a plug which prevents the escape of the spermatozoa. We also know that the queen will fly out more than once, to mate again, if by mischance she does not receive enough sperm to fill her spermatheca (see below); and there is evidence that she will repeat her excursions in a later season.

F. Ruttner has made exhaustive studies of mating from the aspects of both behaviour and anatomy. He has shown (1961) that the oviducts can contain the spermatozoa of some half-dozen drones, and produces evidence to support his opinion that multiple mating is probably a common or even regular event.

The fertile queen

During a few hours after her return to the hive, the spermatozoa pass from the oviducts into the *spermathecal duct*, a small tube inserted in the roof of the vagina, and ascending this duct they arrive at the *spermatheca*, a spherical sac capable of holding about

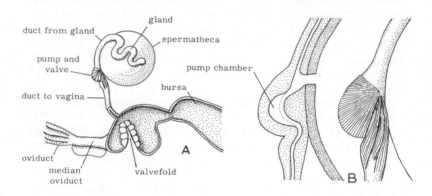

Fig. 33.   A, the spermatheca and vagina, with adjoining organs, of a queen. B, the spermathecal valve and pump, in section, and external view, showing muscles.

7 million sperms (Plate 16 and Fig. 33).   Here they are stored for an indefinite period, until they have been used up. The migration of the spermatozoa is most probably brought about by *chemotaxis*, or automatic response to chemical stimulation, a common biological phenomenon.   The source of the attractive substance may be the *spermathecal gland*, the two branches of which are conspicuously looped over the dorsal surface of the sac. The common duct of these two branches joins the spermathecal duct above the opening of the latter into the spermatheca. The gland is believed to produce a nutrient secretion.

In a useful résumé of earlier work and his own observations, Ruttner (1956) suggests that the migration of sperms into the spermatheca is assisted by muscular contractions of the median oviduct and vagina with simultaneous closing of the valvefold (see below), though it is not clear how this could force sperms into a closed vessel already full of fluid. Below its connection with the sac, the duct embodies a combined valve and pump (Fig. 33): an S-bend in the duct is provided with three sets of muscles which increase the curvature of the bend, decrease it, and compress it; the effect being to draw a small dose of sperm into the S, to close the duct above the sperm, and then to force the sperm down the duct towards the vagina. The queen is thus enabled to release small quantities of semen and to pass these to the vagina as each egg passes the opening of the duct, and also to withhold sperm from eggs which are to be deposited in drone cells.

Just below the opening of the duct, a muscular fold in the floor of the vagina projects upwards as a sort of flap. This may be used to press eggs against the opening of the duct as sperm are released. This is the *valvefold*, a structure of importance in instrumental insemination.

Only eggs laid in worker cells are fertilized; drone eggs, laid in their larger cells, are not. What determines fertilization is not certain, but it is supposed to be the queen's reaction to the diameter of the cells.

Many estimates have been made of the number of eggs laid in a day, and throughout her life, by a good queen. These estimates have been based on observations and counts under a variety of conditions, and the results show great differences. The most reliable are those obtained by photographing combs at regular intervals and counting the numbers of capped cells, but even this method is not above criticism, for the disturbance involved in opening the hive and moving combs must have an effect. Many factors, including that of temperature, affect the rate of laying. Consideration of all these methods, factors, and results leads to the conclusion that an estimated maximum of about 3,000 eggs per day in the height of the breeding season is probably not far wrong. This figure agrees well with the opinions of observant practical bee-keepers. It represents about twice the weight of the queen herself. She probably lays between one million and half a million eggs during her natural lifetime.

The queen is fed, by workers, on brood food alone only when she is laying.

Virgin queens, workers, and laying workers (Plate 17)

In young virgin queens the ovaries are small because the ovarioles, not yet functional, have not expanded. Their spermathecae contain a clear fluid, and often collapse when the insect is opened for dissection. If the queen is not mated the clear fluid solidifies after a variable time, and she is then 'stale'. An unmated queen can lay only drone eggs.

The ovaries of workers are rudimentary, and may easily escape observation when worker bees are dissected. The oviducts are long and narrow, and each bears only a few short ovarioles. The vagina is very small, and there is only a vestigial spermatheca, consisting of a small tongue projecting from the roof of the vagina. Workers cannot copulate with drones, though drones often attempt to couple with them. In colonies which have lost their queens, or when queens begin to fail, laying workers appear. Some of their ovarioles become functional, commonly two or three, or up to six in each ovary, and these produce short strings of eggs. These are laid in both worker and drone cells, sometimes several in a cell, and placed on the sides of the cell as well as its base. Very rarely, a queen is raised in a stock which has no fertilized eggs, a remarkable occasion which has given rise to fairy tales about workers stealing eggs from a queenright stock. The true explanation is that a parthenogenetic egg may, as the result of a simple genetical accident, contain the full, double complement of chromosomes, and will therefore develop into a female insect. Genetical accidents are not unknown. Occasionally a queen, having lost a gene or a number of genes, will produce only white-eyed (and blind) drones, or other monsters. Gynandromorphs, in which part of the body has female characteristics and the remainder male structures, are the result of an accident during the very early stages of the development of the embryo, when a sex chromosome has been accidentally excluded from one of the few cells; all the cells derived from this by subsequent division will then give rise to male structures.

Instrumental insemination

For at least two centuries beekeepers have been trying to control breeding, first by attempting to get queens and drones to mate in enclosed spaces, a method which has always failed, even in large glasshouses. Later, attempts, also unsuccessful, were made to force the coupling of insects held in the fingers. Finally, the injection of spermatozoa by syringe was tried. In 1926 the American Lloyd Watson was the first to achieve limited success with a syringe, and further improvements on the method were made by other American entomologists. In 1944 Dr. H. H. Laidlaw discovered that failure, or only very limited success, was the result of the valvefold in the queen's vagina preventing the semen from flowing into the oviducts. The process of insemination was then greatly improved by using an instrument for depressing the valvefold before introducing the tip of the syringe into the vagina. In natural mating, no doubt, the valvefold is automatically depressed. O. Mackensen and W. C. Roberts (1948) substituted a glass tube for Watson's crude cradle, thus providing the queen with a much more comfortable and unrestricted posture, and Mackensen introduced the use of carbon

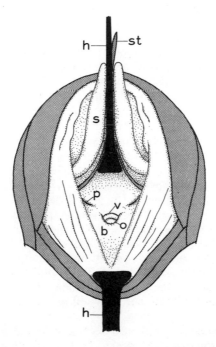

Fig. 34. The end of the queen's abdomen, as seen when prepared for instrumental insemination. *h*, hooks; *s*, sting sheath; *st*, sting shaft; *b*, bursa; *ov*, orifice of vagina; *p*, opening of lateral pouches; *v*, valvefold (after Mackensen and Roberts).

dioxide as an anaesthetic, and incidentally accelerated the onset of laying. Since then, many minor improvements in apparatus and technique have been invented. Semen is collected, with the syringe, from the endophallus in its extreme state of eversion (Plate 15). Fig. 34 shows the open end of the queen's abdomen, with the bursa and valvefold, as prepared for insertion of the syringe.

Obviously, a knowledge of the anatomy of the queen is necessary for expert manipulation, which must, of course, be performed under a dissecting microscope.

Instrumental insemination is not, and is not likely to become, an efficient substitute for the natural process in the commercial or amateur's apiary. Its usefulness is in the field of research in genetics, and in the controlled breeding of improved strains.

The fatherless drone

Eggs which are not fertilized develop into drones. Drones are therefore *haploid*, *i.e.*, their cells contain only one set of chromosomes instead of the usual double set of *diploid* organisms.

In most animals, as well as plants, both sexes are diploid, and when their eggs and sperms are formed a special process, the 'reduction division', takes place, during which a single set of chromosomes goes into each of two cells; consequently each egg or sperm contains only a single set, and is haploid. Fertilization restores the normal diploid condition, the egg then containing chromosomes from each of its parents, who in this way contribute to the genetical make-up of their offspring.

The drone has no father, and has only a single set of chromosomes from his mother. When spermatozoa are being formed in his testes there is no reduction division (in fact there is preparation for a reduction division, but it aborts), and all his sperms contain identical half-sets of chromosomes. The fact that drones receive chromosomes only from their mother does not, however, mean that a queen's sons are all identical. She has a double set, and in the reduction division some of her mother's and some of her father's chromosomes go into each egg, but not the same assortment. Thus, although her drone eggs are not fertilized, her sons will have a variety of different genes.

Her worker daughters will also receive different assortments of their maternal grandparents' genes, but identical sets from their father. This fact alone will show that the workers in a stock may have many different characteristics, but there are still further possibilities of difference, for unless the queen mated only once her spermatheca will contain sperms from more than one drone. The most conspicuous character, which may be observed easily by watching the alighting board of a hive, is the coloured banding of the abdominal segments. This may vary between solid black and solid yellow, with intermediate grades of part black and part yellow on one to all of the abdominal tergites.

CHAPTER 8

THE JUVENILE FORMS

The egg (Plate 20)

The egg is slightly more than $1\frac{1}{2}$ mm. long, and nearly $\frac{1}{3}$ mm. in diameter at its apex; it tapers slightly towards its base, by which it is attached to the floor of the cell. Its ends are rounded, and it is slightly curved. It weighs about 0·13 mg., or about 1/200,000 oz. When it is laid, it is glued to the floor of the cell by a gummy substance, in a horizontal posture, perpendicular to the floor of the cell. Most beekeepers agree that during the next three days the egg sags, finally resting on the floor; but the American worker E. J. DuPraw has recently denied this, and holds that prone eggs do not hatch.

The larva (Plates 18, 19)

In some groups of insects, like the cockroaches and grasshoppers, and the bugs, the newly hatched young already resemble their parents in their anatomy, and their further development is mainly growth in size and the final acquisition of wings. In others, including the bees, there is no resemblance at all: the newly hatched creature is a grub, still in the last stage of embryonic development, and still showing clearly many of the ancestral worm characteristics. After growth in size, the change to the adult form is a sudden one, appropriately called *metamorphosis*.

The larvae of most insects are active, independent animals, moving about in the open air, and therefore well equipped with eyes and legs (though not wings), and finding their own food, like the caterpillars of butterflies and moths. The larvae of bees enjoy much more sheltered lives. They are protected and fed within the nest, and are almost totally inactive. They are blind and without functional legs or means of locomotion, and their mouthparts are such that they can only suck up semi-fluid food. During the larval phase development is virtually limited to increase in size, without progress in organization.

The external anatomy of the larva is shown in Plate 18.

Segmentation is conspicuous throughout the larval body, except in the head, where the consolidation of the six foremost segments was already complete before hatching. There are no constrictions behind the head and thorax. Of the original 21 embryonic segments, two have already disappeared, so the larva has 13 behind the head. The mouthparts retain their embryonic form (cf. Fig. 5). The internal organs also show their segmental origin very clearly: we have already noticed the arrangement of the respiratory and nervous systems of the larva (Figs. 21, 23). The origin of the alimentary canal in three parts is also still conspicuous in the larva (Plate 19), and the proctodeum, though in contact with the mid-gut, is not yet in open connection with it. A simple *oesophageal valve* at the end of the oesophagus prevents back-flow from the ventriculus. There are only four Malpighian tubules, quite slender in the young larva, but increasing in girth as waste accumulates and distends them. Two long, kinked *silk glands* extend through most of the length of the body, and unite in a common duct which opens in the *spinneret* on the labium (actually on the part of the labium which will be the hypopharynx); these glands originate as ingrowths from the labium, and in the adult bee they become the thoracic salivary glands.

The heart is similar in structure to that of the adult insect, but it extends from A9, where its blind end lies, to just behind the head, and is divided into 11 chambers by 10 pairs of ostia. It is clearly visible through the skin of the back. The aorta is thus only a short organ, and it is not closed on its under side, being U-shaped in section. The dorsal diaphragm is similar to that of the imago, but the ventral diaphragm, though well formed in the newly hatched larva, becomes vestigial in later stages.

The fat body is very large, filling most of the space left in the body, except for clear channels which permit circulation, and it is moored by tracheae, but otherwise it is not anchored and floats in the blood. Showing through the transparent skin, it gives the larva its creamy colour. In addition to fat cells and oenocytes like those in the adult, the fat body includes a third kind of cell in which waste accumulates; these are the *urate cells*.

The brain resembles the adult brain in its general form, but the optic lobes are thin plates of nervous tissue, the antennal lobes are not yet developed, though the antennal nerves are laid down. Rudimentary antennal appendages can be seen on the face. Rudimentary legs and wing pouches can be found under the skin of the thoracic segments.

The larva lies on its side on the floor of the cell, in a pool of brood food (and later pollen) fed to it by nurse bees. The spiracles of the upper side are exposed to the air. (When larvae are 'grafted' into queen cups by beekeepers, there is danger of drowning the grubs if enough 'royal jelly' is put in to cover them.) The diet of larvae has been mentioned in Chapter 3.

The reproductive organs are present in larvae in rudimentary form. The ovaries are paired bodies situated near the middle line of the back in A5. They are much larger in mature larvae reared in queen cells (over 2 mm. in length) than in worker larvae, where they are very small. The testes in drone larvae are relatively large, measuring nearly 4 mm. in length, and lie in A4 to A6. Plate 19 shows a testis *in situ*.

Growth of the larva

After hatching, the larva grows very quickly. Insects have no means of increasing the area of their cuticle, once it has been laid down, to accommodate more bulky contents. They must therefore shed and replace it at intervals. The stages of growth are called *stadia*, and the moults are *ecdyses*. The form taken by the larva during a stadium is an *instar*. Regulation of the phases of growth and moulting is performed by endocrine glands.

The moulting process consists of the softening and dissolving of the inner layer of the cuticle; this frees the exocuticle, which sloughs off. The epidermis then secretes new cuticle layers.

The endocrine glands

The *endocrine*, or *ductless glands*, which pour their secretions (*hormones*) *into* the body instead of *out* of it like other glands, constitute a means of biochemical control of some of the normal processes of growth and development, and maintenance. Compared with the highly developed endocrine system of the human body, that of the insects appears to be very simple indeed. As yet, however, our knowledge of it is very incomplete; it may be more extensive than we know, for small inconspicuous groups of cells, difficult to find and study, have been found to give rise to hormones.

Insects possess inherited genes which are responsible for growth and for the characteristics of larva, pupa, and imago. A hormone, *neotenin*, regulates the action of these genes, holding it in check at the appropriate times. Another hormone, *ecdysone*, promotes moulting, thus initiating a new phase of growth and development. Neotenin is produced in large quantities during the larval phase. When its concentration falls below a certain level, the larva pupates, and finally, when the hormone ceases to be produced, the action of the imaginal genes is released.

Neotenin, the inhibiting hormone, is produced by the *corpora allata*. Their name, meaning 'the bodies which have been moved', refers to their origin as ingrowths of the epidermis of H5, and their subsequent migration to their final position above the pharynx, near the brain. Experimental surgical removal of the corpora allata from suitable half-grown insect larvae is at once followed by premature pupation.

These glands persist in the adult insect—where they probably have a new function, perhaps control of general metabolism—as two small round bodies, seen in Plate 13 as they appear in a dissection.

The moulting hormone, ecdysone, is the result of a chain of reactions. A secretion released by a small group of cells in the protocerebrum stimulates the *corpora cardiaca* to become active. These are two small knots of tissue just in front of the corpora allata, on each side of the aorta. They pour their hormone into the blood, and this stimulates larval *thoracic endocrine glands* to produce ecdysone. This indirect stimulation reminds us of the general supervision of our own endocrine glands by the master pituitary body. The thoracic glands are lacking in the adult insect (they must not be confused with the thoracic salivary glands, derived from the larval silk glands), but the corpora cardiaca persist, though their function in the imago is unknown.

The stadia

The consecutive instars, with their durations, and the ecdyses, are listed in Table 1, and illustrated in Plate 20.

TABLE 1
The Stadia after hatching (Worker)

	INSTARS	ECDYSES
Larva	1st (*ca* $\frac{1}{2}$ to $\frac{3}{4}$ day)	——1st
	2nd (*ca* 1 day)	——2nd
	3rd (*ca* 1 day)	——3rd
	4th (*ca* 1 day)	——4th
	5th (*ca* 2 days)	— suppressed
Prepupa	6th (*ca* 2 days)	——5th
Pupa	7th (*ca* 8 days)	——6th
Imago	8th	

The periods indicated in Table 1 are approximate: there is some variation, for the growth process is likely to be affected by fluctuations in temperature, and possibly by other factors.

After the 4th moult, the larva makes the most growth, adding 40 per cent. to its weight. About two days after this moult, the larva receives its last meal, and its cell is sealed. The connection between the ventriculus and the hind-gut is then opened, and at the same time the Malpighian tubules open into the hind-gut; excreta then enter the hind-gut, and are voided in the cell. The adult Malpighian tubules are already being formed. The larva changes its position, and stretches itself out along the cell, on its back, with its head towards the cap of the cell. It spins a cocoon from the secretion of its silk glands, which exudes from the spinneret, finishing this work about 24 hours after the cell is sealed. It then lies motionless in the cell for another day. During this quiescent phase the larva goes through the *prepupal* instar (Plate 20). Although the preceding ecdysis is suppressed, the prepupa is an instar in its own right. This is a period of preparation for the pupal stage. Great and rapid changes in the body take place. The head and mouthparts are remodelled, the thoracic segments alter in shape, and the petiolar constriction forms behind the propodeum (A1), which thus becomes a part of the thorax. The terminal segments of the body are telescoped into A7, and the sting begins to take shape, its initials becoming conspicuous on the ventral surface of A8 and A9 before they disappear within A7. The empty ventriculus is now reduced to a mere narrow ribbon. All these changes take place within the old larval skin.

This complete change of form is *metamorphosis*.

The pupa

Finally, the 5th moult occurs, and the pupa is revealed, with all the external features of the adult insect, except that the wings are not yet expanded, but are small pouches pressed against the sides of the thorax. The pupa is soft and white in colour. During the moult, the linings of the fore and hind-guts, and of the tracheae and glands, are cast off, as well as the larval skin. All the muscles, except those of the heart, break down completely, and the contents of the body are nearly all liquid, with suspended fat cells.

During the 8 days of this instar, the pupa gradually assumes colour, beginning with the eyes, which become pink, then purple, finally brown. The rest of the body passes through deepening shades of tan until it reaches its full adult colour.

The whole of the viscera are recast, except the ovaries and testes, which continue to develop. A new temporary epithelium is laid down for the ventriculus, and the adult stomach is built up on this by budding from numerous points. Its posterior extremity is again plugged until the regeneration of the whole organ is complete.

The last (6th) ecdysis occurs 8 or 9 days after the 5th (4 to 5 days in the queen, $7\frac{1}{2}$ to 8 in the drone).

Having cast off the pupal skin, the young adult bee nibbles away the cap of the cell with its mandibles, and emerges as a small, soft, feeble, greyish creature, as yet unable to sting.

TABLE 2
Life-histories of the Castes

Weeks	Days	QUEEN	WORKER	DRONE
1	1 2 3 4 5 6 7	– – –egg laid– – – – – – hatch – – – – – – – – – – – –	– – –egg laid– – – – – – hatch – – – – –diet changed–	– – –egg laid – – – hatch – –diet changed
2	8 9 10 11 12 13 14	– – – sealing – – – – – – – – – – – – – 5th moult – – – – – – – – –	– – – sealing – – – – – – – – – – 5th moult	– – – sealing
3	15 16 17 18 19 20 21	– – –emerges – – – – – – – – – – – – mature	– – – – – –	– – 5th moult
4	22 23 24 25 26 27 28	⎫ ⎬– – mates ⎭ – – – – – – – – –	– – –emerges – – – – – – – – –	– – – emerges
5	29 30 31 32 33 34 35	– – – – – – – – ⎰– begins to lay	– – – flies	
6	36 37 38 39 40 41 42	– – – – – – – – – – – – – – – – – – – stale if not mated – – – – – – – –	– – – mature – – – – – – – – – foraging begins	– – – mature
8	—	– – – – – – – – –	– – – – – – – –	– – dies, if not already slain
			foraging for 5-6 wks. (rough average)	
12	—	– – – – – – – –	– – – dies	
		dies after several years or is replaced		

PART II
DISSECTION

PART II

DISSECTION

Chapter 9

APPARATUS AND METHODS

1. The Dissecting Microscope

For the study of both external and internal anatomy a microscope is essential. Its cost may seem to present a problem to many students, but we shall see that it is not difficult to improvise an entirely satisfactory substitute for the rather expensive ready made instrument. (It is suggested that every beekeepers' association should obtain at least one dissecting microscope for the use of its members. Its value to the association does not end with the study of anatomy: it is also indispensable for some of the diagnostic operations in connection with bee diseases, and it can be a constant source of delight at apiary meetings for the observation of eggs, brood, emerging bees, etc.)

More disappointment than pleasure and satisfaction is caused by any attempt to dissect with the aid of simple lenses. The strongest lenses which can be used do not give sufficient magnification for our purpose, and their working distances (the distance between the lens and the object) is so small that it is difficult to use tools; also the eye must be closely applied, resulting in a most uncomfortable posture. Thus the worker cannot see what he is doing properly, and is subjected to great eye strain and general discomfort.

The orthodox compound microscope is also unsuitable for dissection. Its lenses reverse and invert the image, causing the movements of tools to appear in the opposite direction to that in which they are in fact moving. In these conditions it is impossible to use scissors and knives with the necessary sureness and skill. This type of microscope is useful only for the examination of slides under high powers. It is essential for some diagnostic work, and useful, if not quite essential, for some aspects of anatomical study (see Appendix II).

The type of instrument required for dissection and for the other uses mentioned above is a *prismatic* microscope, of which three versions are illustrated in Fig. 35. Instruments of this kind embody prisms (on the same principle as those in field glasses) which erect

93

the image, so that we can see our tools moving in the correct directions. Binocular instruments give stereoscopic vision, enabling us to perceive depth in the object as well as to use both eyes. Working distance is ample (at least 3 ins.), permitting the operator to work in comfort. The most useful magnification is about ×20, and higher powers are neither useful nor desirable.

Fig. 35 shows instruments made by firms whose reputation is so high that they need no recommendation. The writer has used Leitz's 'binocular magnifier' for forty years, and Beck's Greenhough model for over twenty years, both with complete satisfaction, and has recently found Baker's 'Sterimag' very good. Both the Leitz and Beck models can be had on a variety of stands, including both of the types illustrated. Inclined eyepieces (shown on the Beck instrument, on which they are optional) are not always an advantage; sometimes they compel an awkward attitude, and the prospective purchaser should try both upright and inclined eyepieces before making his choice. (Inclined eyepieces are most desirable on *large high-power* microscopes.)

Though very fine instruments, and very desirable possessions, these microscopes are expensive. But that consideration need not deter the student, for it is not difficult for a handyman to construct a monocular dissecting microscope at very slight cost, and instructions for this are given in Appendix I.

An indispensable accessory is a 'spot lamp', which will throw a powerful spot of intense light on the insect which is being dissected. Suitable lamps are on the market, but are expensive, and for most purposes home-made lamps are more satisfactory. The writer has made several, and their construction is described in Appendix I. The 'Sterimag' microscope has a built-in lamp, the transformer being housed in the base of the column (but see comments in the Appendix). This instrument, at the time of going to press, costs only £34.

2. Dissecting instruments

Only a few instruments are needed, but it is important that three of them should be of exactly the right kind. The scissors (Fig. 36c) should be 'cuticle' scissors, not less than $3\frac{1}{2}$ ins. in length, and not much longer, with very fine points which cut cleanly right up to their tips. Looked at sideways, they should be *very slim*. The finest small embroidery scissors are much too thick. So far as the writer can discover, the only satisfactory kind comes from France, and they are stamped 'Nogent'. The fine scissors sold to biology students are much too coarse, *i.e.*, too thick in profile. Embroidery

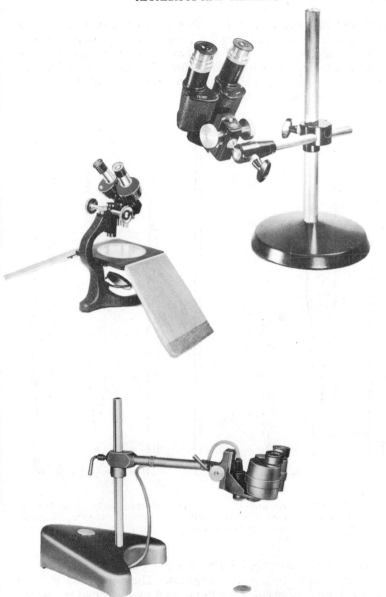

Fig. 35. Dissecting microscopes. *Above:* the binocular magnifier of Messrs. E. Leitz (Instruments) Ltd.; *middle:* one of Messrs. R. and J. Beck's Greenhough models; *below:* the 'Sterimag', with built-in transformer and spot lamp, ×20 model, made by Messrs. C. Baker Instruments Ltd.

scissors can be ground down to reduce the thickness of the blades (taking care to avoid overheating if a carborundum wheel is used). Scissors should be looked after very carefully, thoroughly cleaned and dried after use, and never used for other purposes. The best scissors are exceedingly attractive to womenfolk, and they should therefore be carefully hidden.

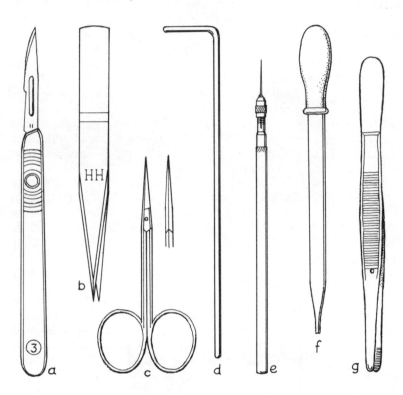

Fig. 36.   Dissecting instruments.   *a*, Swann Morton scalpel with No. 11 blade; *b*, watchmaker's forceps, HH pattern; *c*, cuticle scissors (Nogent); *d*, bent wire; *e*, needle in chuck-top holder; *f*, pipette with teat; *g*, coarse forceps.   Half actual size.

Forceps of the most useful kind are not obtainable from shops which supply to biologists and surgeons. The only really efficient forceps for micro-dissection are those used by watchmakers, which have very fine points and grip very firmly at their extreme tips. There are several patterns, the best for our purpose being marked 'HH'. They are much cheaper than the finest 'biological' forceps, costing at the time of writing only 3*s*. 6*d*., and they can be obtained from E. Gray & Son, 7 Clerkenwell Green, London, E.C.1. (These forceps are ideal for the acarine dissection.) See Fig. 36*b*. A very sharp and finely-pointed knife is the third important tool. This may be a surgeon's eye knife, or the smallest kind of entomological scalpel with an acutely angled point. It must be kept sharp by frequent careful treatment on a fine Arkansas oilstone. For those not accustomed to this rather delicate and skilled job, there is an alternative and equally good knife, the Swann-Morton scalpel, with replaceable blades which can be discarded when they become blunt (Fig. 36*a*). Swann-Morton No. 11 blades have the correct shape; the handle should be No. 3.

A pair of needles, mounted in wooden handles or chuck-top metal handles (Fig. 36*e*) will be in constant use. They can be home-made. Avoid long, whippy needles; the projecting needle should not exceed about ¾ in. in length. Sewing needles of medium thickness are best; they should not be very fine, or they bend too easily. The eye ends should be broken off, to give a convenient length.

Two or three other instruments are required. We need a pipette of the old fountain pen filler kind, with a rubber teat (Fig. 36*f*); a stout wire, bent into an L-shape (Fig. 36*d*), its long limb being about 6 ins. long and the short one ¾ in. long—the best material is brass rod, $\frac{3}{16}$ in. thick; and any old pair of coarse forceps, similar to Fig. 36*g*.

The brass wire has to be heated. If a Bunsen burner or gas ring is not handy, a spirit lamp may be used.

Two or three dissecting dishes are to be made from flat round tins about 3 ins. diameter. These are to be filled with melted beeswax to within ¼ in. of the top of the rim; the wax must then be allowed to solidify. The surface of the wax has to be remelted frequently, and this is done most conveniently by turning a Bunsen flame downwards over the dish. If a Bunsen is not available, a hand gas tool may be used or a butane blow-lamp; otherwise the whole of the wax will have to be melted.

3. Reagents

Alcohol has a variety of uses. The stock should be 95 per cent. *industrial methylated spirit*; in the mystique of the Inland Revenue this strength is known as '66 per cent. overproof'. It can be purchased at small cost out of bond, a permit having been obtained from the local Customs and Excise Officer, who will make no difficulties if the use to which the spirit is to be put is explained. The best dissecting fluid is 30 or 50 per cent. alcohol, made by diluting the stock with water (30 parts spirit to 65 parts water by volume, or 50 to 45, respectively). Water cannot be used by itself because it will not wet the hairy bodies of bees. Alcohol is also the basis of preserving fluids, which will be dealt with in due course, and is much used in making microscope slides. The common dyed methylated spirits of the oil shops, and surgical spirit, should be avoided; they are quite unsuitable.

A few ounces of commercial formalin (38-40 per cent. solution of formaldehyde) will be needed for preserving fluids.

Bees can be killed with many volatile substances. The best is chloroform (solvent quality), obtainable from a pharmacy. Other fluids have disadvantages, and are not worth mentioning.

Glacial acetic acid will be required if Carl's solution is preferred for preservation; it has many advantages.

Glycerine, diluted with water, is an excellent preserving fluid for finished dissections (see section 9 below).

4. Material and its preservation

Freshly killed bees will be available for dissection at all times except in the depths of winter. (Students who do not keep bees should lose no time in making friends with beekeepers; the best way to do this is to join a local association.) With some very important exceptions, however, preserved bees are more satisfactory for dissection, and stocks of these should be built up. In general, the material supplied by biological supply firms is badly preserved and expensive. Black bees, or dark hybrids, are more useful than golden or pale races, the internal sclerotized parts being more deeply coloured and therefore more conspicuous. Bees in a number of different conditions should be preserved, e.g., workers which have been confined to the hive for some time; workers which have just returned with a load of nectar (catch these on the alighting board and kill them at once); young bees just emerging from their cells; winter bees taken early in the spring; laying workers; specimens attacked by acarine, Nosema, or amoeba diseases; larvae and pupae in all stages, though these are best dissected after very short preparation. Queens and drones must be dissected immediately after killing; their reproductive organs become brittle when they are preserved by any method, and thus break up when they are disturbed.

Small samples of bees are killed by slipping into the matchbox in which they are collected a small scrap of blotting paper soaked in chloroform, the box then being set aside for five or ten minutes. A complete colony of bees can be killed with about four teaspoonfuls of chloroform poured on to two pieces of corrugated cardboard, one of which is pushed through the hive entrance and the other placed under the quilt; the entrance is then plugged with a cloth, and the top of the hive blanketed or otherwise sealed to prevent escape of the gas.

Diseased stocks which have in any case to be destroyed, and run-down queenless stocks or nuclei, provide useful sources of material for preserving in bulk. Queenless stocks which have dwindled severely provide large numbers of laying workers during the summer months. In all these cases it is sufficient to stupefy the bees with chloroform and then, before they recover, to immerse them in the pickling fluid.

Alcohol is commonly supposed to be a good preservative. This is a fallacy. It does not penetrate sufficiently rapidly to prevent the onset of decomposition, which ensues very rapidly after death, and it renders tissues very brittle. It is therefore necessary to add to the alcohol some substance which has great penetrating power and which tends to toughen tissues. Two such fluids are commonly used, formalin and acetic acid. Formol alcohol is prepared by mixing the following: 95 per cent. alcohol, 70 parts by volume, formalin 5 parts, water 25 parts. The formalin penetrates quickly, and hardens soft tissues, making them leathery and easy to handle. This fluid can be used not only for preserving material, but also for fixing and hardening finished dissections of freshly killed bees, prior to mounting them in glycerine.

For preserving material, however, the writer prefers Carl's solution. This is made up of 95 per cent. alcohol, 17 parts; formalin, 6 parts; water 28 parts; glacial acetic acid, 2 parts. The acetic acid should be added just before using, i.e., 2 parts of acetic acid to 51 parts of the remaining mixed constituents.

When bees are pickled in bulk, plenty of fluid should be used—about twice the volume of the bees; and the storage jars should be well corked or capped.

Larvae and pupae can be extracted from their cells without damage by floating them. This is quite easy if a jet of water from a fine nozzle is directed into the cells. They should be carefully handled, using a small spoon or section lifter, and at once put into formol alcohol or Carl's solution. They should be dissected two days later.

5. Examination of external structures

With insects, the study of external anatomy is in fact the study of the features of the exoskeleton. Some of the skeletal structures are ingrowths which cannot be seen from the outside, and these are dealt with in the chapters on dissection.

Apart from the microscope and lamp, very little apparatus is required. It is better to use freshly-killed specimens, in good condition, for this part of the work: in preserved bees the hair is matted and the surface dulled, and dried specimens are stiff and brittle. The thick 'fur' obscures the view of much of the body, especially the thorax, and it must be removed. The easiest way is to pour melted paraffin wax (candle wax will do) over the insect, allow it to set, and then peel off the wax, using forceps and needles and the point of the scalpel; the 'fur' comes off with the wax. Alternatively, the hairs can be removed by scraping or rubbing with the side of a knife or a needle. For examination, the specimens can be held in the fingers, impaled on acarine needles, pinned to cork, or glued to cards in convenient postures.

6. Preparation for dissection

Students who have been unable to obtain assistance or advice often get into difficulties through using unsuitable methods. A common fault is to treat an insect in the same way as larger animals like rats or rabbits, pinning it down on its back and opening it in air. Pinning is too clumsy a means of anchorage for small insects: the pins get in the way of the tools and damage the insects. Exposed to air, the viscera cling together in an unrecognizable sticky mass, which quickly dries up. Insects are usually dissected with their backs uppermost, though in some cases another posture may be more appropriate for some special reason. They are fastened down by partially embedding them in melted wax, and they are opened under dissecting fluid, which floats and supports the internal organs and of course prevents them from drying.

For the first attempt, which should be a general dissection, take a freshly killed or preserved worker, selecting one with a well-distended abdomen, and cut off its wings, legs and proboscis with scissors (an old pair of small pointed scissors will do for this rough work). If the bee has been preserved, dry it as well as possible by rolling it gently on blotting paper. Now consult Plate 1. With coarse forceps, sieze the bee by the thorax, back uppermost; with the other hand take the bent wire and heat its short limb in a flame; apply the hot wire to the wax in the middle of the dissecting dish, thus forming a small pool of melted wax somewhat bigger than the bee. Now place the bee quickly in the pool, hold it there with the cool end of the wire, and withdraw the forceps. Reheat the wire and melt a little of the wax near the sides and ends of the bee, and push this melted wax against the body, so that it piles up slightly and makes good contact; this will ensure that the specimen is firmly anchored and will not come adrift during dissection. The bee should be sunk nearly halfway in the wax. Plate 1A shows this operation, and B the embedded bee. All this must be done quickly and without overheating the insect. Dissecting fluid (diluted alcohol) should be poured on at once, enough to cover the bee. The dish may now be placed under the microscope, and is ready for work. Focus the microscope and adjust the spot lamp. Instruments should be laid out ready for use.

After one or two attempts this operation will be found to be very easy and quick to manage, and can then be perfected. A better posture is obtained if, when lowering the bee into the pool of wax, the tip of the abdomen touches the wax first; it will adhere, and then the body can be drawn forward slightly, thus stretching the abdomen a little.

Whole insects, or parts like the head, can be prepared in any posture that may be desired. When the student has had enough experience to wish to preserve his dissections, a variation of the method will facilitate mounting permanently, and this will be described in section 9 of this chapter.

7. Methods of dissection

Details of procedure are described in later chapters, but the general treatment and use of tools may be mentioned here. The first step is to remove the overlying body wall, in order to reveal the internal organs. This is most easily done with scissors in the case of the abdomen, and with the point of the knife when opening the thorax and the head.

Sometimes scissors and forceps are used together, a part being held with the forceps while it is severed with the scissors. The viscera (soft organs) are manipulated, separated, and laid out with the two needles. Forceps are used for removing overlying organs and débris. The pipette is in constant use as an irrigator, for flushing out the body cavity and thus removing fragments which obscure the view.

The movements of the hands and fingers are slight and delicate, and steadiness is essential. A comfortable sitting position should be arranged; a rather low chair is an advantage. On no account should students perch on a high stool at a high laboratory bench. The feet should be on the ground and the knees should be slightly higher than the seat. If the hands begin to tremble, they should be rested. Trembling is the result of strained positions of the hands and wrists, but such strain is easily avoided. The movements of the right hand (in right-handed people) are necessarily limited, and they should be confined within their comfortable range. Thus, when opening the abdomen with scissors (Plate 1C) the operator should not attempt to work round the bee with his right hand, which will at once involve him in awkwardly difficult attitudes, but should move the bee, by revolving the dish with the left hand, thus keeping the right hand in its most comfortable position. The scissors should rest on the rim of the tin dish, and should be steadied against the left thumb. The scissors should be held with the knuckle upwards.

When the knife is in use it may be held like a pen, but with the knuckles downwards, or like a penknife, with the knuckles upwards, according to the work that is being done. When it is used to cut through the body wall (Plate 1D) only the extreme tip of its point is inserted, and the cutting stroke is directed outwards, away from the body; thus damage to the organs just below the body wall is avoided. Similarly, when the scissors are in use, the blade which is inserted should be kept as far as possible parallel to the body wall, and not pointing inwards.

8. 'Semi-sections'

Miss A. D. Betts (1923) has described a method of demonstrating internal anatomy without formal dissection. This consists of slicing through previously hardened bees with a razor, thus exposing certain aspects of the internal organs. The method is of limited value: nothing is so instructive as the usual procedure of removing the body wall neatly and then 'unpacking' the contents. Nevertheless, some useful purposes are served by these 'semi-sections', as Miss Betts called them, particularly those of the thorax, the head, and the abdomen of the queen.

Formol alcohol may be used for hardening, with double the normal addition of formalin. Bees should be left in the fluid for a few weeks. A new safety razor blade is the best knife, the bee being laid on cork and steadied with the fingers.

9. Preservation of dissections

Finished dissections can be preserved in perfect condition indefinitely, and are then useful for reference or for teaching purposes. It is, however, not easy to remove the specimens from the wax in which they have been embedded without damaging them, and therefore, when the student has acquired enough skill to be confident of making a good job, it is more convenient to prepare the insect, in the first place, by a method which facilitates its subsequent mounting.

This method involves mounting the bee, before dissection, on a small sheet of mica, or cellulose acetate, or $\frac{1}{32}$ in. perspex, which is cut to fit the vessel in which the dissection is to be kept. Instead of wax, a mixture of equal parts of beeswax and resin, melted together, is used. Some of this melted compound is spread on the sheet of mica or plastic, and its adhesion is ensured by piercing a number of small holes in the sheet, to provide a 'key'. The beeswax-resin mixture holds the specimen much more firmly than plain wax. It is melted with the hot wire, and the bee is attached, just as it is in the dissecting dish, except that it is not necessary to embed so deeply. The whole assembly is then laid in the dissecting dish, and the hot wire is applied to the dish wax near the corners of the small plate, thus anchoring it; then the dissecting fluid is poured on, and the work is begun. When the dissection is completed, it is only necessary to detach the corners from the dish wax with a penknife, and the dissection, on its plate, can be lifted out without disturbance.

The bee, if it is a freshly killed insect, should now be soaked for a few hours in 10 per cent. formalin in water or dilute alcohol to harden it and prevent subsequent swelling. A pickled bee does not need this treatment.

The preservative fluid for finished dissections is a simple one: it is composed of equal parts of glycerine and water; after mixing, this fluid is to be brought to the boil to expel dissolved air; boiling should not be continued, but the liquid should be allowed to cool, and it should be kept in a tightly stoppered bottle. If air is not removed from solution, it will appear later as thousands of minute bubbles adhering to the specimen, and these are difficult to get rid of. Pure glycerine is a good medium, but its higher refractive index causes some delicate structures to become virtually invisible. The equal-parts mixture is, in this respect, just right for our purpose: it causes some structures to appear semi-transparent, but not to disappear.

The next stage in preparation is to soak the dissection in some of the glycerine solution in a small dish, leaving it for two or three days. The solution then displaces the dissecting fluid, and the large quantities of air in the tracheae of freshly killed bees will be dissolved. Finally the dissection is ready to be placed in fresh solution in the vessel in which it is to be kept.

This vessel may be a small specimen tube, but cylindrical vessels are not satisfactory, the curved sides acting like lenses and causing optical distortion. The old-fashioned method of sealing dissections in deep cells on microscope slides is unsatisfactory because the pressure set up by expansion in rising temperatures eventually (sometimes quickly) fractures the sealing cement. Such preparations are notoriously short-lived. What we want must be regarded as small museum specimens rather than as microscopical preparations, and we must adopt a technique suitable for them.

The best form of container is a kind of miniature museum jar, the 'micro-jar', the construction of which was originally published by the author in the *Journal of the Quekett Microscopical Club*, and is now described in Appendix I. These containers are made of perspex, with lids which are attached by screws.

Before filling, the jars should be wetted inside with alcohol; shake this out and dry the mouth of the jar. Spread petroleum jelly, or burette cock lubricant (paraffin-rubber, or silicone grease, obtainable in tubes or tins) over the mouth and the underside of the lid. Fill the jar with glycerine solution, and insert the specimen. The jar will now be over-full, the fluid standing up above the mouth; put on the lid, thus squeezing out the excess fluid and enclosing no bubbles; then screw down the lid. Finally, wash off the exuded glycerine solution under the tap, and wipe the jar dry.

Fluctuating temperatures may cause slight leakage of the fluid and its replacement by air bubbles. The air will be held by the bubble trap, and thus prevented from wandering over the dissection. If it becomes too large, the lid can be taken off, a little fresh fluid added, and the lid replaced. Other instructions for maintenance are given in the appendix.

10. Preservation of dry specimens

Specimens arranged to display the external anatomy can be preserved in glycerine solution like dissections, but dry mounts are equally, or indeed more, satisfactory. Their preparation involves merely glueing them with seccotine or some similar adhesive to cards and covering them to protect them from damage and dust. The guiding principle is on no account to attempt to *seal* them under glass, for the underside of the glass is very likely to become etched and obscured. They should be ventilated, but kept in a dry place. A convenient form of mount, the size of a microscope slide (3 by 1 ins.), is described in Appendix I. These preparations are very liable to be spoiled by mites, small insects, and moulds, and to prevent this they should be poisoned by painting the undersides, which do not show, with a solution of mercuric chloride in alcohol. Storage in a dry place is very important. Most living rooms are much damper than is supposed. The best plan is to keep these mounts in a tightly closed box together with some activated silica gel.

11. Plan of work

The arrangement of the succeeding chapters follows roughly the order in which the student will pursue his studies. There can be no hard and fast distinction between external and internal anatomy, but in practice it is convenient to divide work roughly on those lines. It is suggested that the external structures should be studied first (Chapter 10), until they are quite familiar.

Then, turning to dissection in the stricter sense, the student should begin by making several experimental general dissections of the worker, using both preserved and freshly-killed bees. In the course of this work he will gain familiarity with his tools and methods of manipulation, as well as with the lay-out of the internal structures. He will then be able to make some careful preparations for permanent preservation. Next he may embark on some special dissections, including the demonstration of a few features of the worker which are not well displayed by approach from the dorsal aspect. All this work is covered in Chapter 11.

The head requires special treatment and dissection from different aspects, and forms the subject of Chapter 12.

The drone and queen (Chapters 13 and 14) are of interest primarily on account of their remarkable reproductive apparatus, but also because they differ from the worker in many of their structures. The supply of queens is limited, and the student will wish to have acquired some skill by practice before he attempts to dissect them. Experience and confidence can be obtained by dissecting workers during the months of winter and spring, before the outdoor work makes large demands on the beekeeper's time, and before drones and queens become available.

The juvenile forms are more difficult to handle, and will probably be tackled last of all (Chapter 15).

CHAPTER 10

EXTERNAL ANATOMY

(For notes on method, see Chapter 9, section 5)

1. Worker, head and mouthparts (Plates 2, 3)

Lay the bee on its side, and examine the attachment of the head to the thorax, noting the articulations with the pleurites of T1, which 'float' on the neck membrane. Then turn the insect back upwards, and find the ocelli on the vertex.

Anterior aspect. Identify the regions and organs flagged in Plate 2C. Note the articulations of the mandibles, labrum, and antennae, moving these with needles.

Posterior aspect. Identify the regions and organs from Plate 2D.

Proboscis. Remove the head. Hold it down, or pin it down on a cork with an acarine pin (double needle) inserted astride the foramen. See how the proboscis swings on its cardines: place the point of a needle behind the postmentum (Plate 3) and push the proboscis into the extended position.

Remove the entire proboscis: push it into the extended position, look into the fossa, and see where the cardines are articulated to the upturned walls of the fossa; detach them with the point of a knife; the whole apparatus will come away easily (first attempts may result in damage to the cardines). Put the proboscis into water in a watch glass or small dish, and notice that the post- and prementum, the stipites, and all members behind them are connected by a sheet of membrane, continuation of the floor of the mouth cavity. The proboscis can be spread out flat by tearing this membrane with needles. (If the operation has been successful, the proboscis can now be prepared for mounting as a microscope slide: see Appendix II.)

The action of the proboscis can be watched under the dissecting microscope if a few bees are placed in a small box covered with a piece of glass, small spots of honey or syrup being placed on the side of the box near its rim.

2. Worker, thorax and appendages (Plates 4 to 7)

Examine the thorax from all aspects (Plates 4, 5), rubbing off the hair, when this is necessary, with the end of a needle or the knife. Identify the regions and parts flagged in the Plates. The minute second spiracle can be seen only in a newly emerged bee, or in an advanced pupa. Notice particularly the tergite of the first segment, with its lobe covering the first spiracle, a feature of importance in the acarine dissection. Observe the articulations of the wings with the thorax. Notice the scutal fissure, which permits distortion of the thorax by the indirect wing muscles. The large spiracle of the propodeum is conspicuous. (The internal features of the exoskeleton of the thorax are dealt with in the next chapter.)

The wings (Plate 6). Details of the articular sclerites and hamuli can only be seen with higher powers of the microscope. Venation is of interest to the taxonomist and geneticist, but not particularly to others.

The legs (Plates 6, 7). Remove the legs, holding them gently with forceps and cutting the intersegmental membrane above the coxae with the point of the knife. Identify the segments and organs shown in the Plates, especially the antenna cleaner on the fore-legs, the tibial spine on the middle legs, and the pollen combs (scopae), tibio-tarsal joint (pollen press), and the corbiculae of the hind legs. Details of all these, and of the foot, are more clearly seen under higher powers (Appendix II).

The use of the arolium can be watched under the dissecting microscope if a few bees are placed in a small box covered with a piece of glass. The arolium comes into action when the bees walk upside down on the glass, and it is then clearly visible.

3. Worker, abdomen (Plate 4)
Identify the segments and organs flagged in the Plate. Detach some of the tergites and sternites and examine them with higher powers of the microscope to see the spiracles on all the visible segments, as well as on the spiracle plate of A8, also the 'wax mirrors' and the canal of the scent gland on A7 (Plate 12).

4. Drone, compared with worker
The external anatomy is similar to that of the worker, with the following differences:

Head (Plate 2). The compound eyes are larger, and meet at the vertex, forcing the ocelli forwards on to the frons. The antennae are stouter, and have one more segment than the worker's. To see the sense organs, make a microscope slide (Appendix II). The mandibles are small and toothed, and the proboscis is shorter than the worker's.

Thorax. This is larger than the worker's, to accommodate more powerful flight muscles, but is otherwise similar. The wings are larger than the worker's. The hind legs have no special apparatus for collecting and packing pollen.

Abdomen. The abdomen is broader and blunter than the worker's. Its tip, A8, is partly visible, with two claspers on each side (homologous with the worker's sting stylet and sheath). There are no wax mirrors and no scent gland.

5. Queen, compared with worker

The queen is very like the worker, with the following exceptions:

Head. This is very slightly smaller than the worker's. The compound eyes are somewhat smaller, the proboscis is shorter. The mandibles are larger, and serrated (Plates 2, 3).

Thorax. This is similar to that of the worker. There is no special apparatus for pollen collection and packing on the hind legs, but their outline resembles that of the worker more than that of the drone.

Abdomen. The abdomen is longer than the worker's, each tergite and sternite being correspondingly longer. The overall length is not so apparent in a virgin queen, whose ovaries are not yet expanded. There are no wax glands.

6. The juvenile forms (Plates 18, 20)

Extract larvae, prepupae, and pupae from brood combs (see Chapter 9, section 4). Some of the larvae and prepupae may be hardened in formol-alcohol. For some purposes it is better to stain them in erythrosin; this makes it possible to see more clearly some of the surface features which are otherwise difficult to see on account of the lack of contrast in the colourless, transparent skin backed by masses of white fat cells. Or most of the fat can be removed by puncturing the skin of a fresh specimen and soaking it in acetone, afterwards washing out the acetone with alcohol and mounting in glycerine.

Larva. Examine a large coiled larva, and identify the features flagged in Plate 18. Cut up a hardened specimen and examine the face and the appendages (sting initials) of A8 and A9. Drone larvae have only the appendages of A9 (initials of the claspers).

Prepupa. Note the changing mouthparts, developing antennae, and rudimentary wings; the legs are also more advanced than in the larva. All these features show up better in a specimen which has been soaked in pure glycerine. The initials of sting or claspers are larger.

Pupa. Examine a pupa which has reached the tan-coloured stage. All the parts except the wings, which are small pouches folded against the sides of the thorax, are already in their adult forms, but they are not yet hairy, and are very easily seen. Note the surrounding envelope of pupal skin.

7. Preserved specimens

Preserve the following: Dry-cell preparations: heads of all castes, mounted to show all aspects, fur removed. Thorax of worker, showing lateral aspect, wings and fur removed. Abdomina, dorsal and ventral aspects. Legs of worker, the hind legs to show both inner and outer surfaces. Mandibles, all castes. In glycerol, stained larvae, prepupae, pupae, of workers and drones. See Appendix II for micro-preparations.

G

CHAPTER 11

DISSECTION OF THE WORKER BEE

1. The general dissection

Plates 8 to 10 illustrate the dissection of the whole body, from the dorsal aspect, in three stages. In practice it will be found convenient, especially for the inexperienced student, to begin with the abdomen, and to complete its examination before starting on the thorax and head, and the directions which follow are arranged on that plan.

For the study of the alimentary canal, the heart, the tracheal sacs, and the ovaries in both normal and laying workers, bees should be dissected immediately after killing. These organs become brittle in preserved specimens. For all other purposes, preserved bees are much more satisfactory. Bees with fully distended abdomina will show the heart and nervous system to the best advantage.

The specimen is prepared, fixed in the dissecting dish, back uppermost, and covered with dissecting fluid, as described in Chapter 9, section 6, and illustrated in Plate 1. The dish is in position under the dissecting microscope, and instruments are laid out ready for use. This book, with its plates unfolded, should be propped up on a book rest and protected from spilled fluid!

Beginners should not be discouraged by initial failures: everyone makes an unholy mess of the first few attempts.

2. The abdomen

Turn the dish so that the head of the insect points to 10 o'clock. During operations, remember to move the dish to suit the convenience of the right hand and scissors (Chapter 9, Section 7).

Stage 1: exposing the viscera (Plates 1, and 8, A and B)

Open the abdomen: steady the dish with the left hand, and the scissors by resting them against the left thumb on the edge of the dish (Plate 1). Insert one point of the scissors under the overlapping edge of the tergite of A5, on the right side of the body. Cut through the body wall, and continue snipping through the right side, working forwards towards the thorax. Keep the inner blade of the scissors as far as possible parallel with the side of the insect, thus avoiding thrusting it in deeply and damaging the viscera.

110

When the corner at the front end of the abdomen is reached, turn the dish clockwise to suit the scissors hand and cut across the broad front of the abdomen to the opposite corner. Then again turn the dish and work down the left side. When the tergite of A6 is reached, turn the dish again and cut across the tergite, taking great care not to damage the soft organs underneath, and so complete the circuit at the beginning of the first incision.

The roof of the abdomen is now to be lifted off; it should be free and easily removed. Lift it gently with the point of a needle. If it resists, one or more of the infolded parts of the tergites have not been severed. These uncut parts must be found and cut through, using the inner scissors point as a probe while gently lifting the roof with the needle. When it is clear that the roof is free, do not pull it off roughly, but lift it gently with two needles. There may be slight resistance from tracheae, but these will break without doing any damage. If, however, it seems that the internal organs are being pulled out or disturbed, take the roof by its edge with the fine forceps, and with the needle in the other hand break the tracheae, which will show as fine threads stretching between the roof and the organs below. Finally lift off the roof and turn it over. If the work has been done neatly, it will come off in one piece.

Examine the underside of the roof (Plate 8C) as it lies in the dissecting fluid. Observe the heart, with its closed posterior chamber and its ostioles (there are five pairs, but the anterior pair may have been lost), in the mid-line of the roof; also the dorsal diaphragm, transparent, but clearly visible, and its attachments to the apodemes of the tergites. Note also the pericardial fat cells, large numbers of small, creamy bodies clustered against the heart, and the dorsal sheet of the fat body forming a pad between the heart and the body wall. Some of the abdominal muscles may also be seen as flat, nearly transparent bands stretched across the tergites. In preserved bees the heart and dorsal diaphragm occasionally adhere to the viscera and thus tear away from the roof. Having examined all these organs, lay aside the roof, and look at the contents of the abdomen.

The appearance of the undisturbed viscera is very variable, depending on the state of the alimentary canal (Plate 8, A and B). In a bee which has been confined to the hive for some time, or a young bee which has not yet flown, the rectum is greatly distended by accumulated faeces, the bulk of which are yellow pollen husks (A). If the rectum has been damaged by instruments during the opening operations, some of the faeces will have escaped, and will litter the dissection. If the bee has just returned to the hive after a flight, the rectum will be empty and shrunken to very small proportions; if she has brought home a load of nectar or water, the crop (honey stomach) will be expanded into a large, transparent globe (B); if it is empty it will appear as a small, semi-opaque, pear-shaped body. (The student should obtain bees in all these conditions for dissection.)

Part at least of the ventriculus will be visible as a broad, corrugated tube. A loop of the small intestine will be found connected to the forward end of the rectum; its other end, which joins the ventriculus, may not be visible. The slender, tangled threads which spread all over the abdomen are the Malpighian tubules.

In a freshly killed bee, the tracheal sacs will be seen as large bags, silvery with included air (which escapes when a needle point is inserted), obscuring parts of the other organs. In preserved bees the sacs are almost invisible, filmy membranes, the air having been dissolved by the preserving fluid; this is an advantage, for the very large and numerous sacs, when filled with air, are confusing to the beginner. However, a little experience of work on both fresh and pickled bees will clear up this small difficulty. When air-filled sacs obscure the view, they should be pulled out with forceps. Tracheae in large numbers appear as silvery tubules in all parts of the body. From time to time during dissection it will be necessary to clear away débris (faeces, fragments of tissues, etc.) which collects in the abdominal cavity. This is done by irrigation with clear dissecting fluid, a jet of which is directed into the cavity with the pipette.

Stage 2: displaying the alimentary canal (Plate 9)

Take a needle in each hand, pass them under the rectum and ventriculus, and lift up the alimentary canal, gently tearing away the network of investing tracheae in which it hangs, and lay it over to the left side, as shown in the Plate. A little careful teasing out of the tracheae and Malpighian tubules will permit the canal to lie loosely, showing all its parts. Identify and examine the parts. Notice the six rectal pads, which appear as whitish bars on the wall of the rectum. The small intestine is a narrow coiled tube with six longitudinal pleats. At its junction with the ventriculus, about one hundred Malpighian tubules are inserted. This is the pyloric region of the canal. Food masses in course of digestion can usually be seen in the ventriculus, showing as dark areas where the corrugations of the ventriculus are smoothed out. If the ventriculus is torn with a needle, this food mass will exude as a brownish gelatinous substance. If the crop is full of nectar, the proventriculus will be visible through its walls. In any case, tear open the wall of the crop, using needles, and turn up the proventriculus so that its four triangular lips may be clearly seen. If they are closed, the lips meet to form a cross. If they are partly open, an aperture like a four-pointed star is seen. The forward end of the crop narrows into the oesophagus, which enters the thorax through the petiole.

Stage 3: The underlying organs (Plate 10)

Grasp the alimentary canal with forceps, and stretch it. Cut through the oesophagus with scissors, and treat the rectum in the same way, cutting through it as far back as possible. Carefully cut away the remaining small triangle of the roof which was left at the tip of the abdomen, also lower the side walls with scissors, giving a better view of the floor. Flush out the cavity with the pipette. Compare with Plate 10.

The ventral diaphragm may not be noticed at first, but closer inspection will reveal it as a transparent film which very slightly obscures the view of the chain of ganglia and other underlying features. In a later dissection, the diaphragm may be studied more carefully; it is attached to the apodemes of the sternites; its anterior end extends into the thorax and is attached to the furca of T2 and T3, while its posterior end is anchored to the spiracle plate of A8. The diaphragm should now be torn out with the fine forceps, taking care not to damage other structures in the process. The organs lying on the floor of the abdomen will then be more clearly visible. The most conspicuous is the chain of five ganglia, connected by twin longitudinal commissures. The last, the 7th ganglion, is attached to the sting apparatus, and comes away with the latter when it is torn out of the worker's body after stinging. The main lateral nerves which spring from the ganglia can be seen running out to right and left; those of the 7th may be seen passing to the muscles of the sting.

The fat body spreads widely over the floor of the abdomen, being particularly well developed over the wax glands of the sternites of A4 to A7. Smaller clusters of fat cells occur along the sides of the abdomen. The fat body is highly developed in young bees and winter bees, where the cells are large and plump, but in old foragers they are shrunken. The abdominal muscles show clearly, some of the larger sets being very conspicuous as broad V-shaped pairs of bands stretching between the thickened forward margins of adjacent sternites.

It is unlikely that the ovaries will be seen in early attempts at dissection. They are difficult to see, and since they encircle the alimentary canal they are torn away when it is lifted out. In a later dissection they can be found as follows: after removing the roof of the abdomen, lift the alimentary canal slightly from the right-hand side, and look sideways under it, when the right ovary with its oviduct will be seen as an almost transparent, narrow, flat tube running to the root of the sting. Gently disengage the ovary from the tracheae which tie it down and attach it to the other viscera. Repeat this operation from the other side, thus freeing the left ovary. Go on with the dissection, removing the canal. The ovaries will then be seen lying or floating in the abdominal cavity, their oviducts disappearing behind the sting, their distal ends separated. In the undisturbed abdomen, the tips of the ovaries are joined and attached to the heart (Plate 17).

The sting, if not wholly visible, can be examined *in situ* by removing more of the wall of the abdomen at the tip (Plate 10). Identify the parts flagged in Plate 11, dorsal aspect. The whole apparatus can be removed intact very easily by passing needles below it and lifting it out, the small muscles which hold it giving way without offering noticeable resistance. The extracted apparatus can now be turned over, as it lies in clear fluid, and its ventral aspect (Plate 11) can be examined.

Very rarely, the sting apparatus is laterally reversed, the only evidence of this being that the positions of the venom gland and the alkaline gland are reversed.

The powerful muscles of the sting apparatus conceal the plates which constitute the system of levers actuating the lancets. The plates can be exposed by removing the muscles by maceration (see Appendix II). Note that the sting apparatus is arched; it can be flattened by tearing away the proctiger, which is firmly attached to the oblong plates, and it is then easier to examine and also to mount as a microscopical preparation.

3. The thorax

In a freshly killed bee it is not possible to see the structure of the indirect wing muscles, or to remove them without great difficulty, for they are extremely soft and elastic. In bees which have been preserved in Carl's solution or formol-alcohol, however, these great muscles are hardened and tough, very easy to examine and to handle. Therefore preserved bees should always be used for work on the thorax.

Stage 1. Exposing the flight muscles (Plates 8A and 12)

The roof of the thorax is best taken off with the knife: insert the extreme point only, as shown in Plate 1, and then, by an outward and forward stroke, make a short slit in the body wall; continue this along the dotted line in the Plate, all round the domed roof of

T2; then make a longitudinal slit along the mid-line. Usually the roof is very firmly attached to the flight muscles, and must be detached, again with the point of the knife. Keep the blade in a horizontal position, pass its point under the body wall, through the longitudinal slit, and separate the roof from the muscles by small forward movements, gradually working the point further under the body wall. When the first half of the roof is nearly free, steady it with forceps while completing the separation. If this is done carefully, the muscles will be undisturbed and undamaged, and will have the appearance shown in Plate 8. Now remove the other half. Finally take off the remainder of the roof, along the second dotted line in Plate 1. This is not attached to the muscles, and will come off easily. If necessary, remove more of the side walls of the thorax, down to the level of the wings. The indirect flight muscles are now exposed.

Stage 2. Oesophagus and glands (Plate 9)
Remove the flight muscles: simply grip bunches of them with the forceps and pull them out (this is virtually impossible in a freshly-killed bee). Observe the attachment of the longitudinal muscles to the 2nd phragma, which is an extension inwards of the tergite of the second segment. (These attachments show very clearly in a median longitudinal 'semi-section'.)
Below the longitudinal muscles observe the oesophagus, passing from abdomen to head, and the salivary glands of the thorax (derived from the silk glands of the larva). The aorta is a delicate tube which is destroyed by the removal of the indirect flight muscles; it can be found by careful lateral dissection.

Stage 3. The nervous system in the thorax (Plate 10)
With forceps remove the oesophagus and salivary glands. The combined furcae of T2 and T3 are now conspicuous, forming a strong strut and bridge across the thorax and protecting the great second ganglion, which can be seen below it. The first ganglion lies in front of and partly concealed by the furca of T1, at the anterior end of the thorax; to see this it will probably be necessary to remove more of the body wall in this region. Both ganglia can be exposed by removing the overlying parts of the furcae, using the point of the knife and forceps. Sheets of semi-transparent muscles attached to the furca will also have to be removed; this must be done cautiously, for the ganglia are very easily damaged. Notice the thick twin commissures joining the ganglia. Commissures also run forward from the first ganglion to the suboesophageal in the head, and from the second ganglion to the abdomen, where they

join the abdominal chain. Nerves from the second ganglion can also be seen passing to the propodeum and into the abdomen, where they serve A2.

4. The head (Plates, 8, 9, 10)

Dissection of the head from the dorsal aspect is not the most informative approach, but should not be omitted. (See Chapter 12.)

Stage 1.   Exposing the brain, etc.

Using the point of the knife, cut along the dotted line of Plate 1C. The isolated portion of the wall will come away easily. Conspicuous on the summit of the brain are the purple retinae of the three ocelli. Extend the opening to each side, cutting away part of the compound eyes (Plate 8A); both knife and scissors can be used for this part of the work. In a freshly killed bee the brain is obscured by a large tracheal sac filled with air, and this must be cleared away; the protocerebrum, bearing the ocelli, will then be revealed, also the optic lobes connected with the compound eyes. The pigmented parts of the compound eyes indicate the radiating ommatidia. In front of the brain the hypopharyngeal glands are conspicuous in young nurse bees and winter bees, in which the acini are large and white, but in foragers they are much shrunken and may be difficult to find. Behind the brain will be found another branch of the hypopharyngeals, and a small part of the postcerebral glands may be seen.

Stage 2.   Displaying the glands and antennal lobes (Plates 9, 10)

With the point of a needle, lift out the glands, noting the form of the hypopharyngeals, like strings of onions, and the quite different branched structure of the postcerebrals.

Cut away the frons down to the level of the antennae, when the antennal lobes, with nerves running into the antennae, can be seen if the head is lying in a favourable position, as in Plate 10.

5.   Lateral dissection of the worker

Mount a preserved worker on its right side.

The head

The dissection is chiefly useful to demonstrate the suspension of the proboscis, but also shows different aspects of the organs which are the principal subjects of other dissections.

Take off the left gena, and carefully work downwards, lowering the walls. Note the tentoria, two powerful struts running from the foramen to the clypeus, and joined at their feet by a small transverse bridging piece. The upturned wall of the fossa lies just below the gena, and the end of the cardo will be found articulated to a knob on the edge of this wall. The oesophagus enters the head through the foramen and then expands into the pharynx. The long extensions of the hypopharyngeal plate embrace the pharynx. The cibarium is connected to the clypeus by its dilator muscles. (The organs revealed by this dissection will be recognized from the diagrams in Figs. 4 and 15.)

The thorax

Remove the left side of the thorax and excavate, removing the flight muscles and lowering the walls as the work proceeds, observing: the attachment of the vertical muscles to the floor and roof of the thorax; the attachment of the longitudinal muscles to the roof and first phragma in front, and to the second phragma in the rear. When approaching the mid-line of the thorax, try to find the aorta, between the right and left sets of longitudinal muscles, and below it, the oesophagus. See the ganglia and commissures and try to see the anterior end of the ventral diaphragm, which enters the thorax through the petiole. Finally, having removed everything else, observe the direct wing muscles on the inner surface of the right hand wall of the thorax. A special preparation to show the direct muscles can easily be made by slicing off one side of a thorax with a razor blade and trimming away débris.

'Semi-sections' of the thorax, both transverse and longitudinal, show some of the structures very clearly and diagrammatically, and are worth preserving (see Chapter 9, section 8).

The abdomen

Examine the sting chamber. First cut a window in the side of the abdomen, starting at the rear edge of the tergite of A6; enlarge the opening towards the posterior end of the abdomen, severing the muscles which hold the isolated parts of the body wall by passing the point of the knife under them. The spiracle plate of A8 will be found overlying the sting apparatus, and having something of the appearance of a dog's head (Plate 5). The spiracle plate is bound to the quadrate plate near the dog's nose. Note the proctiger, attached to the sting apparatus and also to the rectum. Remove the contents of the abdomen, lower the walls, and survey the abdominal muscles of the right side (some of them are shown in Fig. 11).

6. Dissection of the laying worker (Plate 17)

See note on obtaining material in Chapter 9, section 4. Laying workers should be dissected when freshly killed, or within two or three days after pickling, or the ovaries will become too brittle to handle.

Remove the roof of the abdomen. If the crop is full and the ovaries are well developed, the anterior extremities of the ovaries may be seen crossing the crop, but this is not usual. Proceed as though dissecting out the ovaries of a normal worker (Chapter 11, stage 3). The laying worker's ovaries are much more easily seen, as the ovarioles contain strings of eggs. With ordinary care, after breaking the binding tracheae, the alimentary canal can be removed without tearing the ovaries apart, and they are then displayed as shown in Plate 17.

7. Preserve the following:

General dissection, freshly-killed worker, viscera undisturbed, rectum full; ditto, rectum empty, crop full; general dissection, freshly killed bee, alimentary canal displayed (stage 2); roof of the abdomen from a bee with distended abdomen, showing the heart and dorsal diaphragm, etc.; preserved bee with distended abdomen, showing the nervous system, sting *in situ*, etc. (stage 3); the sting removed, all aspects; laying worker; lateral dissections to show head with suspension of the proboscis and internal organs, indirect flight muscles (also transverse and longitudinal 'semi-sections' of thorax); the direct wing muscles; abdominal muscles; the internal features of the exoskeleton (furcae, apodemes, phragmas, etc.); any interesting abnormalities, *e.g.*, laterally reversed sting, wax-bound bee, gynandromorph; the sting chamber; and if the student is interested in pathology, the acarine dissection, two bees side by side in one jar (see the writer's *Laboratory Diagnosis*, 1949).

CHAPTER 12

DISSECTION OF THE HEAD, ALL CASTES

1. Preparation

Heads are removed and mounted for dissection from both anterior and posterior aspects. Beeswax alone is too soft to hold the head firmly, and a stiffer cement must be used. The wax-resin mixture (Chapter 9, section 6) is suitable, and may be used to mount the heads on small slips of mica, acetate, or perspex, these in turn being temporarily anchored in the dissecting dish by melting the wax in the dish at their corners with the toe of the bent wire. The heads should be embedded deeply, with their upper surfaces almost level with the surface of the cement.

2. Worker's head, from the anterior aspect (Plate 13)

Stage 1. The glands

Use the point of the knife to cut through the wall of the mask, across the vertex, round the margins of the compound eyes, and round the edges of the mask, excluding the mandibles, labrum, and clypeus. The antennae should be snipped off near to their insertion. There is now nothing to stop the mask from being lifted off except the tentoria. Cut round the small pits in the suture surrounding the clypeus. The mask will then lift off. The clypeus is held down firmly by the cibarial muscles (dilators of the cibarium), and these must be disengaged by using the point of the knife in the same way as when taking the roof off the thorax. In a young bee, five or six days old, or in a winter bee which has not yet nursed brood, the hypopharyngeal glands will have the appearance shown in the Plate, the acini being plump and creamy white; they will almost fill the space in front of the brain, as well as sending branches to the back of the brain. They can be lifted out to show their string-of-onions structure. In foraging bees which have completed their nursing duties, these glands are greatly shrunken, almost to the point of disappearance, leaving only thin thread-like remains. Notice also the mandibular glands.

Stage 2. The brain

The hypopharyngeal glands having been removed, the brain is exposed. Identify the structures flagged in the plate. Cut down the compound eyes to the level of the optic lobes, and note the indications of the radiating ommatidia. Remove the roof of the cibarium and examine the floor of the mouth cavity, finding the hypopharynx, with the two pores marking the ends of the ducts of the brood-food glands.

3. Worker's head, from the posterior aspect (Plate 13)
Stage 1. The glands

Remove the proboscis. Then remove the wall of the occiput and postgenae, by the same means as in the preceding dissection. Clear the tentoria in the same way, and after lifting off the back wall of the head cut down the tentoria, to give a clear view. Identify the structures flagged in the plate. The postcerebral glands are more translucent than the hypopharyngeals; note their branching form.

Stage 2. The brain and suboesophageal ganglion

Remove the glands. The cut end of the oesophagus will be seen projecting from the space between the brain and the suboesophageal ganglion. If the corpora allata are not visible, pull the oesophagus out a little way with forceps, when they should come into view.

As more skill is acquired, the principal nerves (Fig. 24) and the ducts of the glands (Fig. 15) can be traced.

4. The head of the drone

Proceed as with the worker. The hypopharyngeal glands are absent, and the postcerebrals are reduced to rudiments adhering to the posterior wall of the head. The mandibular glands are also vestigial. Clusters of fat cells in some parts of the head may be mistaken for glands.

The brain proper (part of the protocerebrum under the ocelli) is smaller than the worker's, though the very large optic lobes give a misleading appearance of size.

5. The head of the queen

The mandibular glands (source of queen substance) are much larger than the worker's. The hypopharyngeal glands are lacking (vestiges of their ducts may sometimes be found). The postcerebrals are like the worker's. The brain is somewhat smaller than that of the worker.

6. Preserve the following:

Anterior and posterior aspects of the worker's head; stages 1 and 2 of each can be mounted in the same jar. Similar specimens of the heads of drone and queen for comparison. (Lateral dissections of the head to show the suspension of the proboscis were included in the list at the end of Chapter 11.)

CHAPTER 13

DISSECTION OF THE DRONE

1. Preparation for dissection

Immature drones which have just emerged from their cells, or which can be caught actually emerging, should be killed and dissected at once. Mature drones cannot be killed without partial eversion of the endophallus ensuing. Partly matured drones, if their development has not proceeded too far, can usually be killed without inducing eversion, though ejaculation often begins. Pickled material is useless for dissection of the reproductive organs, but preserved drones may be used for the dissection of the head, of the thorax to show the powerful flight muscles, and of the abdomen to show the immensely powerful abdominal muscles.

2. Immature drone, dissection from the dorsal aspect (Plate 14)

Stage 1. Exposing the viscera

Proceed, as in the case of the worker, to take off the roof of the abdomen. In crossing over at the anterior end it will be difficult to avoid damage to the testes, which will then be slightly frayed out in that region. Notice the resistance to the scissors offered by the greatly developed abdominal muscles. In all dissections of the drone great care must be taken to avoid damage to the mucus glands; if these are pierced, mucus will exude, coagulate, and obscure the work.

Compare the undisturbed viscera with the plate and identify the organs. Note the enormous testes, composed of bundles of tubules which can be shown by teasing out one of the testes.

Stage 2. The reproductive apparatus

Lay out the testes, as shown in the plate (where only one testis is drawn). If they are not very gently handled, they will break off at the vasa deferentia. Remove the alimentary canal: grasp the ventriculus with forceps and draw it out backwards until the crop appears; cut through the oesophagus with scissors, also through the rectum. Flush out the abdominal cavity with clean dissecting fluid from the pipette. This exposes the rest of the reproductive apparatus. Identify the parts flagged in Plate 14.

The whole of the apparatus can be removed for more detailed examination by cutting through the body wall, round the genital aperture, with the point of the knife. It can then be lifted out, very gently to avoid damage, fixed in formol-alcohol, and preserved.

3. Maturing drone (Plate 15)

Treat as in the preceding dissection. The testes are reduced in size, finally becoming thin, triangular, translucent scales. The seminal vesicles, having received spermatozoa, are increased in size, so are the mucus glands. If eversion has not occurred, there may yet be partial ejaculation, the evidence of which may be seen in the swelling of the ejaculatory duct near the bulb.

4. Mature drone

Catch a flying drone on the alighting board of the hive, and kill it. Cut off the everted endophallus and mount the insect for dissection in the usual way. After removing the roof of the abdomen, look for the testes, now shrunken to mere yellowish or greenish scales. In the second stage of the dissection, compare the condition of the rest of the apparatus (except the endophallus, which is now outside the body) with that of the immature and maturing drone. The ejaculatory duct will be seen passing out through the genital aperture.

5. The everted endophallus (Plate 15)

Kill a mature drone, examine the partially everted endophallus, and compare it with the drawing in Plate 15. Now take the abdomen between the thumb and forefinger and compress it laterally; eversion will continue until the bulb reaches the tip of the organ, when sperm and mucus will be liberated.

Handle living drones, noticing how easily eversion is induced by the slightest pressure, and the 'snap' when this occurs.

6. Preserve the following:

Immature drone, stages 1 and 2; mature and maturing drones, stages 1 and 2; the reproductive apparatus, removed from a maturing drone; abdomina of drones, showing partially and completely everted endophalluses.

CHAPTER 14

DISSECTION OF THE QUEEN

Queens must be dissected immediately after killing. If they are preserved the ovaries become extremely brittle and cannot be handled for proper examination. Their heads may be pickled, however, and can then be dissected at leisure (Chapter 12, section 4). Some queens, however, should be pickled in strong formol-alcohol for semi-sections (Chapter 9, section 8).

1. Dissection of the fertile queen (Plate 16)

Stage 1. Exposing the viscera (Plate 16A).

Mount the queen for dissection in the usual way, dorsal surface uppermost. Open the abdomen carefully, avoiding damage to the underlying organs when taking off the roof. When the roof appears to be quite free, raise the front edge of it slightly with the forceps, and with a needle held in the right hand gently detach the soft organs which cling to the roof. If this is not done very carefully the ovaries will be pulled out and spoiled. The undisturbed viscera will then have the appearance shown in Plate 16A. At their anterior ends, the ovarioles of the two ovaries are joined, and at this point were attached to the heart, from which they were detached before lifting the roof. Identify the parts flagged in the plate, and note that the enormous ovaries fill about two-thirds of the abdomen. The rectum is always empty.

Stage 2. Displaying the reproductive organs (Plate 16B)

Remove the alimentary canal: grip the ventriculus with forceps and stretch it, drawing it towards the rear end of the abdomen; cut the canal in front of the crop and through the rectum. This must be done without damaging the ovaries.

Very carefully slip a needle under one of the ovaries and lift it, turning it slightly outwards; the tracheae which bind and suspend the ovary will now be seen stretched; break them with a needle held in the other hand, until the ovary is freed sufficiently to lie over to the side, as shown in the plate. Repeat this operation with the other ovary.

Lower the sides of the body wall near the tip of the abdomen. Flush out the body cavity with the pipette. Note the paired oviducts,

123

which converge to the root of the sting, where they join the median oviduct, the spermatheca, with its gland and investment of silvery tracheae, and the acid gland of the sting, which is much longer than the worker's. The sting apparatus is firmly anchored and cannot be lifted out as it can from a worker dissection. The shaft of the sting is curved, not straight like the worker's.

Queens used for dissection are usually discarded ones, a year or two old, sometimes older. In these old queens the fat body and Malpighian tubules are discoloured by the accumulation of waste products (this can be seen in the pericardial fat cells; it is not necessary to spoil a good dissection by removing the viscera to expose the fat cells on the floor of the abdomen).

2. Dissection of the fertile queen from the ventral aspect

This more difficult operation will demonstrate that part of the reproductive tract which was concealed by the sting apparatus in the preceding dissection. It should not be attempted until the student has acquired considerable skill and experience of dissection.

This part of the tract, including the median oviduct, vagina, and bursa, together with the sting apparatus which lies above them, are firmly attached to the floor of the abdomen, from which they can be detached only by careful, patient use of the point of the knife.

Mount the insect with the ventral surface upwards. First, using scissors, cut away the floor of the abdomen, starting at the anterior end, and working backwards, removing the sternites until the lateral oviducts are uncovered. Continue cutting through the body wall at the extreme sides of the abdomen, towards the tip. Now, gripping the floor with forceps, at the forward edge, insert the point of the knife and very carefully separate the sternites from the soft parts beneath them. Remove the floor piecemeal, as it becomes possible to cut away parts of it to facilitate the work.

When the dissection is finished, it will be possible easily to identify the bursa and lateral pouches; the latter, in this position of the queen, are uppermost. The remaining structures are more difficult to identify, but will become clearer after soaking in glycerine.

3. 'Semi-section' (Plate 16C)

Prepare queens as described in Chapter 9, section 8. The median longitudinal section is the only useful one. Since queen material must not be wasted, the student should practise on workers; it is by no means easy to cut exactly through the middle of the body. Mount the half-body on its side. Remove the alimentary canal to expose the remaining ovary. The part of the tract under the sting will not be so clearly distinguishable as the plate may suggest, but after soaking in glycerine the parts can be made out fairly well. The depressions in the hardened ovary, shown in the diagram, indicate the space occupied by the alimentary canal.

4. Dissection of the virgin queen (Plate 17)
Dissect from the dorsal aspect, and compare with the fertile queen. The ovaries are small and undeveloped, the ovarioles containing no eggs. The spermatheca contains fluid, and collapses easily. Compare also with the laying worker.

5. Spermatozoa in the spermatheca
To obtain spermatozoa or check their presence, anaesthetize the queen, and kill her by crushing the thorax. Dissect quickly in 0·15 per cent. salt solution. Remove the spermatheca and put into a drop of the salt solution in a watch glass or on a microscope slide. Tear open the spermatheca with needles. If spermatozoa are present they will emerge as a creamy mass with something of the appearance of a tuft of cotton wool; spread this out with needles and examine under a 16 mm. objective or higher power. See Fig. 30. (Permanent preparations: Appendix II.)

6. Queen wasp, queen hornet, queen bumble bee
All these are worth dissecting for comparison with the honeybee queen. The relatively small numbers of ovarioles are characteristic of these less specialized insects. Hornets and wasps are not so easy to dissect as bees on account of the large masses of fat which obscure the viscera.

7. Preserve the following:
Fertile queens dissected from the dorsal aspect, stages 1 and 2. Fertile queen dissected from the ventral aspect. Semi-section. Virgin queen, and queens of wasps, hornets, bumble bees. A 'scrub' queen, for comparison with a queen of good quality. Queens of different ages, to show comparative discoloration of the fat body. The sting.

CHAPTER 15

DISSECTION OF THE JUVENILE FORMS

1. Dissection of the larva (Plate 19)

Choose large coiled larvae. In the fresh state they are too fragile to handle, while if they are fully hardened the fat cells form a solid mass which cannot be broken up without destroying the viscera. They must therefore be hardened by immersion in formol-alcohol for only two or three days before dissection. In any case, the removal of the fat is a rather tiresome process, requiring care and patience to take it away piecemeal.

The coiled larva cannot be straightened out, and must therefore be dissected from the side. Use scissors to remove the body wall. Disengage the fat in small fragments. If this is done successfully, the structures shown in Plate 19 will be revealed. The plate shows a drone larva, more convenient for the work on account of its larger size. The ovaries are very small in worker larvae, much larger in queen larvae; they occupy the same position as the testes, which are still larger.

2. Dissection of the prepupa

Prepupae show the same structures as larvae, but after the faeces are discharged the ventriculus is reduced to a narrow, flat strap, hardly recognizable as the same organ as the enormously distended ventriculus of the feeding larva.

3. Pupae in the early stages may be dissected like larvae, but from the dorsal aspect. Older pupae can be treated like adult insects.

Newly hatched larvae can be mounted whole as microscopical preparations (see Appendix II).

(More experienced students may like to follow Nelson's methods (1924). He stained larvae before dissection, and was thus able to see the internal organs more clearly. His method was to fix in Carnoy's fluid (absolute alcohol 6 vols., chloroform 3 vols., glacial acetic acid 1 vol.), and after washing to stain in Mayer's carmalum overnight, finally destaining in acid alcohol for 4 to 6 hours. Alternatively, Carl's solution may be used for fixing and erythrosin for staining.)

APPENDIX I

CONSTRUCTION OF EQUIPMENT

Construction of a Monocular Dissecting Microscope
Making a binocular prismatic microscope involves obtaining paired lenses and the difficulties of combining converging tubes and adjustments for inter-pupillary distance. This is not beyond the

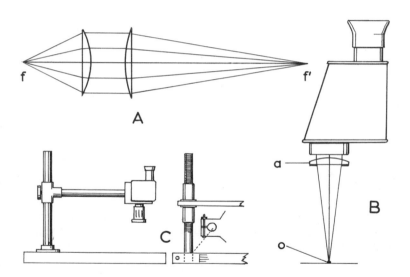

Fig. 37. A, the principle of the 'magnifier' or accessory lens. Two lenses placed near to one another; the points *f* and *f'* are their foci. The image of an object placed at *f* will be focused at *f'*, and *vice versa*. The rays passing between the lenses are parallel. B, the principle applied to the construction of a dissecting microscope: rays entering the field glass are parallel when the object *o* is at the focus of the lens *a*. C, pillar stands constructed from gas barrel and gas fittings.

skilled mechanic, who will require no further instructions; but many students will prefer a simpler job, and it is quite easy to make a monocular instrument which is perfectly adequate for the work. It is true that a binocular gives the advantage of stereoscopic vision, with appreciation of depth in the object, but the advantage is often over-emphasized.

For the monocular microscope, all that is required is a pair of field glasses, which need not be damaged and can continue to be used for their original purpose, or half of an old pair of such glasses, which can be permanently converted, and a small additional lens costing only two or three shillings. A brief account of the interesting optical principle of the system will put the amateur constructor 'in the picture'. A camera set to 'infinity' is adjusted to focus distant objects, and the rays of light coming from distant objects reach the camera in parallel bundles. Accessory lenses, or 'magnifiers', are supplied to put in front of the camera lens to enable close-up objects to be focused and photographed. An accessory lens of 3 feet focal length will feed parallel rays into the camera when the object is 3 feet distant from the camera. These rays will be focused on the film by the camera lens, producing a larger picture than the camera could normally take. Hence the term 'magnifier'. A field glass is a telescope, designed for the examination of distant objects and therefore to receive parallel rays. Making use of the 'magnifier principle' we can convert the field glass into a microscope by putting in front of its field lens an accessory lens which will feed parallel rays into the glass when the object is in focus.

For our present purpose, it is convenient to have a working distance between the front lens of the glass and the dissecting dish of about 3 ins., so this should be the focal length of the accessory lens. With a 3 in. lens a field glass of power ×6 will give the best magnification, i.e., ×18 to ×20. A field glass of power ×8 will require an accessory lens of 4 in. focal length. Field glasses of higher power are not desirable, for the greater working distances involved raise the instrument to an inconvenient height above the bench. If no field glass is available, a prismatic monocular made entirely in plastic can be obtained for 25s. from Combined Optical Industries Ltd. of Bath Road, Slough, Bucks. This can be used with a 3 in. accessory lens. It has a long focusing sleeve which covers a vertical distance of over 1·5 cm., a considerable advantage. The plastic lenses of this instrument are easily scratched, and should therefore be treated with care and protected from dust. Probably the best way to clean them is to swab them gently with soapy water, rinse in distilled water, and allow to dry without rubbing with a cloth. This can be done easily by holding the glass upside down.

The accessory lens should be plano-convex (rounded on one side and flat on the other), and it should be attached to the field glass with its rounded side next to, and close to, the field lens of the glass (Fig. 37B). A plano-convex lens thus placed reduces spherical aberration (an inherent defect of single lenses) to negligible proportions. It can be attached, temporarily, with plasticine, or mounted in a wooden or cardboard holder which can be pushed on to the tube of the field lens; or if the maker has the necessary skills a metal adaptor can be made with clamping screws or some other device. In any case, the field glass need not be damaged.

To obtain the necessary depth of focus it will be necessary to stop down the accessory lens by sticking to it a disc of black paper with a small central hole. Stopping down reduces the amount of light which enters the microscope, so the hole should not be too small. The best compromise will be found by trial and error; about ¼ in. diameter will be about right.

The microscope will have to be mounted on a stand which will hold it firmly over the work. This can be a wooden bridge, or the field glass can be clamped to a retort stand. The best kind of stand, however, is that in which a horizontal arm carrying the microscope is clamped to a vertical pillar set in a steady base, and this can be constructed cheaply and easily (perhaps with the help of a plumber) from standard gas piping and gas fittings, as shown in Fig. 38C. In one of these sketches a 9 in. pillar of ¾ in. gas barrel (¾ in. is its *internal* diameter) is fixed in a batten holder screwed to a stout board base about a foot square. A horizontal arm of ¾ in. piping is screwed into a T-junction which will slide on the pillar and which can be clamped to it by a bolt. The arm should be about 8 or 9 ins. long. An alternative plan is shown in the other sketch. Here the pillar is threaded, with two screw collars and two washers which hold firmly a horizontal hardwood batten and permit its height above the base to be adjusted. In this sketch the pillar is held in a slotted hole in the base, tightened by means of a ¼ in. Whitworth bolt and nut.

A suitable wooden cradle to carry the microscope can be fixed to the end of the arm. The C.O.I.L. monocular, which is very light, is very suitable; it is shown in the first sketch. If the height is adjusted approximately, the focusing device of the field glass will provide sufficient movement to focus accurately on the work.

Those with metal-working skills will most probably make more ambitious plans. The writer (1951) has described the conversion of a discarded field glass body, retaining its prisms but removing its lenses and substituting a microscope objective and eyepiece. This entails fitting a 1 in. brass tube to the body, to carry the eyepiece, and an objective fitting (which can be made from a Beck transport cap). A 2 in. Beck objective and a ×6 eyepiece are suitable lenses.

H

The C.O.I.L. monocular has been converted in this way by N. Milns (1960), and the present author has converted other instruments (1960a); the last-named account concerns Wrays's miniature 'Panora' field glass, the resulting microscope being a pocket-sized portable. Accessory lenses of suitable diameter and focal length can be obtained from Gowllands Ltd., Morland Road, Croydon, Surrey, for two or three shillings.

(The same principle can be employed to convert a field glass or small telescope into a short-range telescope for watching insects at a little distance, e.g., bees on combs, or on the alighting board of a

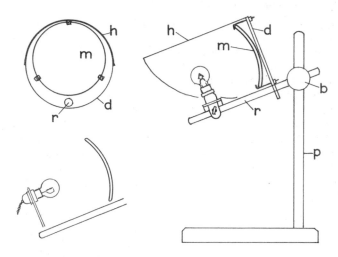

Fig. 38.   General plan of construction of a spot lamp.   r, ⅜ in. brass rod; d, brass disc sweated to r; m, concave mirror, about 3 in. diam.; h, hood sweated to d; b, boss clamping r to pillar p of stand. The lamp holder slides to focus, on r. The left-hand lower sketch shows the arrangement for a straight filament lamp.

hive, or feeding from a syrup container. An accessory lens of about 10 in. focal length is recommended; it need not be stopped down. The enlarged image thus obtained greatly facilitates observation of the action of the proboscis, exposure of the scent glands, oviposition, etc.)

Construction of a Spot Lamp

Domestic lighting lamps with extended filaments are quite unsuitable. We require small lamps with condensed filaments, the cheapest and most robust kind being car headlamps, worked from small transformers with an output of 6 or 12 volts and 5 amps. The writer uses 6 volt 24 watt lamps with V-filaments (Mazda No. 79). V-filaments are not always easy to obtain, but are more satisfactory than the more common straight filament type. However, the latter can be used if necessary.

Either a lens or a concave mirror can be used to throw an enlarged and somewhat out-of-focus image of the filament on to the work. The advantage of the concave mirror is that it greatly simplifies the construction of the lamp-house. Only a shade to protect the eyes from the direct light is necessary, and full ventilation of the lamp is ensured. Fig. 38 illustrates a design used by the writer, and requires no further explanation. Suitable mirrors are obtainable from surplus disposal shops, and cost 2s. 6d. or 3s. The method of mounting V- and straight filament lamps should be noted. Some means of tilting the lamp should be built in: the beam should strike the work at an angle of about 45°. Some dissecting microscopes have built-in spot lamps fitted to throw an almost vertical beam down on to the work; this kind of illumination creates no shadows and thus fails to reveal structure in relief. The built-in lamp of the 'Sterimag' microscope has this defect, but it is understood that the makers intend to provide an alternative fitting. If a microscope with a pillar stand is used, the lamp can be attached to the pillar instead of to a separate stand, preferably on the left side.

If a lens is used instead of a mirror, it should have a focal length of about 2 ins. and a diameter of $1\frac{1}{2}$ ins. or more. The most convenient method of focusing is to mount the lens in a sliding tubular fitting on the front of the lamp-house; the latter should be adequately ventilated.

'Micro-jars' for Preserved Dissections

'Micro-jars' (see Chapter 9, section 9) are easily made from perspex sheet, the component parts being sawn out with a piercing or coping saw with metal-cutting blades. The saw should be used slowly, to avoid generating heat which would melt the plastic and cause the saw to seize up. A form suitable for bees is illustrated (Fig. 39), its dimensions being shown. It consists of two rectangular panes, $\frac{1}{8}$ in. thick, cemented to a U-shaped middle piece; this middle piece must be $\frac{3}{16}$ in. thick for dissections of workers, but $\frac{1}{4}$ in. for queens and drones. A $\frac{1}{8}$ in. lid closes the vessel. The cement is a solution of perspex in chloroform (solvent quality), of syrupy consistency. The assembled parts are to be screwed into a vice or cramp and left overnight to allow the cement to harden. The rough edges of the jars are then filed flat, then rubbed down on the finest glass-paper, which is laid on a piece of plate glass. Finally the whole is polished with Brasso, using a cloth laid on the plate glass. Particular care must be taken with the mouth of the jar, which should be truly flat to make a good joint with the lid. Bubble traps are small areas filed away (as at *b* in the drawings) in the panes, to catch bubbles and prevent them from wandering over the surface of the dissection during examination; they should be filed before the parts are assembled. Holes are drilled as shown, and tapped for 6 BA screws, with which the lid is secured.

If the jars are frequently handled, the panes may become scratched.

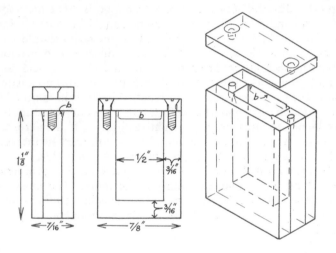

Fig. 39.   Perspex 'micro-jar', actual size, showing dimensions.   The U-shaped middle piece is ¾ in. thick, *i.e.*, the size for worker dissections.

Scratches are easily removed by polishing with Brasso; they may be prevented by cementing large cover-glasses to the front pane with Canada balsam dissolved in chloroform. Labels can be attached to the backs or sides of jars with an adhesive which will stick to plastics, such as 'U-hu', and the labels can be protected by a coat of colourless nail lacquer.

The jars are laid flat on their backs for examination under the microscope, but it is as well to store them in the upright position, to make sure of keeping any bubbles at the tops and thus prevent them from becoming entangled with the specimens. If many dissections are to be stored, it is worth while to make a special box for them. If this is painted inside with cellulose paint, no harm results from possible slight leakage of glycerin.

Model of the scutal fissure

A simple model can be made from one half of a small rubber ball, preferably one with a thick wall, to represent the dome of T2. A slit cut in each side represents the scutal fissure, as in Fig. 8, C and D. When the model is placed on the table and pressure is applied to

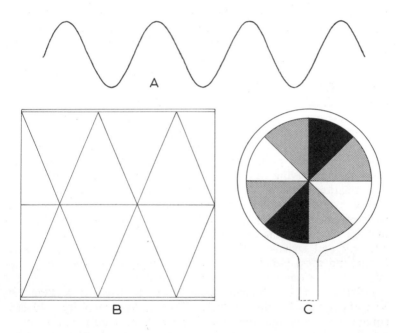

Fig. 40.   Von Frisch's 'artificial eye'.   A, form of a sine wave; B, pattern for polaroid triangles; C, triangles arranged on a perspex holder, and showing an extinction pattern.

its top, the slit will open and permit the model to be distorted in the same way as the dome is distorted by the down pull of the vertical indirect wing muscles. When the pressure is relaxed, the rubber half sphere will resume its original form, and the slit will close, just as the scutal fissure closes when the longitudinal indirect muscles contract.

Polarized light and the Von Frisch 'Artificial eye'

Light waves are 'sine waves'. Fig. 40A shows the characteristic form of a sine wave. The wave shown in the figure is vibrating in the plane of the paper. Waves coming from any source of light vibrate in all directions. If they are passed through a certain kind of crystal, the waves which emerge all vibrate in parallel planes, and are then said to be polarized. Polaroid consists of a large sheet of crystals, sandwiched between acetate sheets to protect it. When light passes through the polaroid sheet, the waves which emerge are polarized, all of them vibrating parallel to the 'axis of polarization' of the polaroid. If we put two sheets of polaroid together, with their axes of polarization at right angles, when they are said to be 'crossed', all the light waves are absorbed, and none can pass.

The von Frisch 'artificial eye' represents the eight retinula cells which are grouped round the rhabdom of an ommatidium. It consists of eight triangles of polaroid with their axes of polarization radiating from the centre of the circle (or tangential to its circumference—either arrangement will do). If we hold this model cross-section eye up towards blue sky, the triangles will allow different proportions of light to pass, and we shall be able to see a sort of pattern, resembling that of Fig. 40C. This is because the sun's light, reflected from any point of the sky, is partly polarized, the direction of vibration varying in different quarters of the sky, according to their relation to the position of the sun. Therefore, if we turn round slowly, while still holding up the model and looking through it, we shall see that the pattern changes. The retinulae of the ommatidia of some insects and crustacea appear to be similarly affected, and signals (which can be electrically detected in large compound eyes) indicating the varying degrees of effect are transmitted by the ommatidial nerves to the brain.

Polaroid can be obtained from H. S. B. Meakin & Sons, 36 Victoria Street, London, S.W.1. The smallest quantity sold is a square 2 in. × 2 in., and the cheaper ('industrial') quality costs 4s. The model can be made from a 2 in. square, which should be placed over the diagram, Fig. 40B. Mark the intersections by pushing a pin through the polaroid, then join these points by needle scratches, and cut out the triangles with scissors. The best way to mount the triangles is to arrange them on a disc of perspex, anchoring them with tiny spots of 'U-hu' at their corners. The polaroid can be protected by covering it with another sheet of perspex, or acetate sheet. It is a convenience to cut the perspex disc with a handle (as in the figure, 40C).

A simple addition to the model makes it possible to demonstrate the effect indoors and at night, or in dull weather. This is a plain disc of polaroid, the same size as the circle of triangles, and pivoted on the same centre by means of a pin, rivet, or small BA bolt and nut. Looking through the model, and revolving the additional disc, causes the appearance of the same effects as we get when we look through the model at the sky and turn slowly round. Similarly, the bee, changing direction in flight, perceives the change of pattern, and can select that which will lead her to the source of nectar or back to her hive.

Working model of the sting

Snodgrass (1933) suggested making a cardboard model of the sting, showing the movements of the plates and lancets. A more durable model in plywood has been found useful for teaching purposes (Fig. 41). A plywood baseboard on two battens (b, b) carries the model of one side of the sting. The fixed, stationary parts (the oblong plate, ramus, bulb and stylet) are cut out from plywood with a fretsaw and tacked and glued to the baseboard; they are shown shaded in the figure. Another fixed piece, p, added at the bottom, completes a groove in which the lancet slides; this is also shaded. The quadrate and triangular plates are cut out of plywood and are articulated as follows. A small round-headed screw goes through a hole in one corner of the triangular plate and is screwed into the oblong plate at x. The opposite corner of the triangular plate carries, on its underside, a small brass plate, drilled with a hole for another screw in the underside of the quadrate at y; a small washer is placed between the quadrate and the head of the screw. The two moving plates slide over some of the fixed parts, and are thus at a higher level; a piece of plywood is glued under the quadrate to raise it to this level and to keep it working smoothly; into this piece is driven a round-headed screw, z, which passes through the slot s in the base, and beneath the base also goes through a brass plate at the end of the lath w. The other end of the lath lies in a channel in the batten at the right-hand side of the base, and projects beyond the base.

The lancet is made from 'non-sag' flexible, spiral wire, curtain rod. It is attached at one end to the triangular plate with binding wire, and lies in the groove formed by the bulb and stylet and the additional piece p. At its other end a brass serrated point represents the barbed point of the lancet.

The moving parts are painted black and the fixed parts of the sting grey or tan. The baseboard and the rest of the model are painted white. Finally, a strip of perspex is screwed to p to prevent the lancet from leaving the groove. Alternatively, two inconspicuous wire bridges across the groove will serve the same purpose.

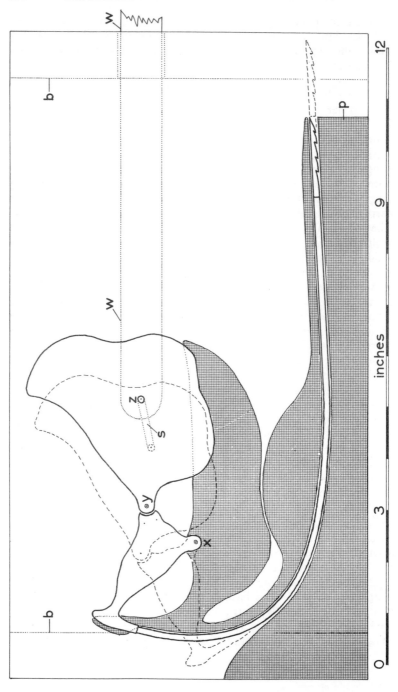

When the lath is pushed in, from right to left, the quadrate moves with it, causes the triangular plate to rotate on x, and the lancet is protracted. In the drawing, broken lines indicate the protracted positions of the moving parts.

Slides for dry mounts

Prepare some hardwood slips (mahogany is a good wood for the purpose) $\frac{3}{16}$ in. thick and exactly the same size as your glass microscope slides (which are not always *exactly* 3 in. $\times$ 1 in.). With a piercing saw cut out circular holes $\frac{3}{4}$ in. in diameter in the middle of the slips, or rectangular holes $\frac{7}{8}$ in. $\times$ $\frac{3}{4}$ in. if these are preferred. Glue a piece of white or grey card over one surface of the slip; this will be the bottom of the slide. Take a piece of thin card 3 in. wide and lay the wooden slide across it. Score the card along the sides of the slide, turn up the sides, and cut them off level with the top of a glass slide laid on top of the wood. Now use black gummed paper or passepartout paper to attach the card to the glass slide, with a $\frac{1}{8}$ in. overlap on the glass. The wooden slide can now be pushed out of its cover, like a match box.

Mount the specimen in the cell, gluing it down with seccotine or some similar adhesive, and label the wooden slip like a microscope slide. Return the slip to its cover, when the specimen and labels will be visible through the glass. It can be taken out again at any time if the glass needs cleaning or if any other adjustment is necessary.

Fig. 41.   Working model of the sting.   Fixed parts shaded, moving parts, in retracted positions, in thick outlines; protracted positions indicated by broken lines.   Half actual size.

APPENDIX II

MICROSCOPICAL PREPARATIONS

A dissecting microscope of the kind described in Chapter 9, section 1, and in Appendix I is indispensable for the study of anatomy, as well as being a most valuable tool in almost every branch of study of natural history. Almost the entire field of bee anatomy can be covered with its use, as perusal of Part II of this book will indicate, and no student should be deterred from undertaking this work by having no access to a microscope of higher power. Nevertheless, higher powers, for which dissecting microscopes are not suitable, do enable the student to see some of the smaller anatomical details, such as have been mentioned in Part I; and it is hoped that readers who can do so will also continue their studies into the field of histology.

An elaborate and expensive microscope is not at all necessary. An ordinary 'students' stand', which can be bought secondhand (with the help of someone who understands microscopes!) is all that is required. It should have 16 mm. ($\frac{2}{3}$ in.) and 4 mm. ($\frac{1}{6}$ in.) objectives, ×6, ×10, and if possible ×15 eyepieces, and an Abbe condenser. The lamp for use with this instrument is simply an ordinary domestic bulb of the 'silverlite' type (not 'clear' or 'pearl'), 40 or 60 watts. (Students who later on work on pathology will require a 2 mm. ($\frac{1}{12}$ in.) oil-immersion objective, in addition.)

A few simple microscope slides are of real use, and their preparation is described below. They fall into two classes: (i) parts of the exoskeleton and its appendages; (ii) the soft internal organs. Quite easy techniques of preparation are adequate. (Ambitious students who wish to attempt microtome sections of organs should consult a paper by F. F. H. Boxall (1956) on the paraffin method. This is by far the best and simplest account of the technique for bee students.)

(1) Parts of the exoskeleton

It is usually necessary to bleach the deeply-coloured body wall partially, to render it sufficiently transparent. Sometimes this can be avoided by taking the material from pupae which have not yet acquired the full, adult coloration. After bleaching, it is usually desirable to remove the soft tissues (muscles, etc.) which are attached to the sclerotized parts.

Bleaching. The bleaching agent is chlorine, which may be applied in various ways, the simplest being to immerse the material in one of the common domestic bleaching fluids. A useful vessel for this operation is a small, shallow, white glass cosmetic jar. The process must be watched carefully, for the action of the bleach may be rather rapid, and there may then be danger of carrying it too far. Some colour should be retained, or the specimen will be invisible when it is finally mounted. Bleaching should be stopped when the material becomes of a medium tan colour, by replacing the fluid with water, and changing the water several times.

Maceration. After bleaching and washing, transfer the specimen to a 10 per cent. solution of caustic potash (potassium hydroxide). Leave it in this for a day or two, or until the soft parts have been dissolved. The process may be hastened by heating, but there is then some danger of damage to the more delicate hard parts, and the slow method is better. Finally, wash thoroughly in several changes of water. Washing should be protracted in cases where the soft tissues are enclosed, as they are in legs. It is suggested that the material should stay for half an hour in each of six changes of water.

If maceration is carried out before bleaching, bubbles will very probably be trapped in closed organs like legs or antennae, and they are then very difficult to get rid of.

(The stock solution of caustic potash should be kept in a bottle with a rubber cork, not a glass stopper.)

Dehydration. All water must now be eliminated. Soak the material in two or three changes of 95 per cent. alcohol (industrial methylated spirit 66 o.p.; not the common dyed methylated spirit). Closed organs, e.g., legs, require longer treatment: about twenty minutes in each change. This treatment is conveniently applied in small vessels, a watch glass or small specimen tube, only small quantities of spirit being used. Finally, pass through two changes of cellosolve to remove the last traces of water. (Cellosolve is a cheap and untaxed substitute for absolute alcohol.)

Clearing. This treatment removes the cellosolve and renders the specimen transparent. Transfer the material to a few drops of clove oil. When it is quite transparent (as seen under the microscope), clearing is complete. Dark areas or cloudiness in closed organs indicate that dehydration has not been complete, and the material should be returned to cellosolve.

Mounting. Wash out the clove oil with xylene ('xylol'). If the clove oil is not removed, the finished mount will become brown. Put a drop of Canada balsam (supplied dissolved in xylene) on a slide, and transfer the specimen, wet with xylene, to it. Arrange the specimen with needles. Warm the slide, warm also a clean cover glass, and with forceps lower the cover glass on to the balsam. Experience will teach the student how much balsam to use to fill the space under the cover glass without too much excess balsam running out over the slide. It is a mistake to use large cover glasses; use as small a cover as possible—a $\frac{5}{8}$ in. or $\frac{1}{2}$ in. cover is big enough for most objects. Apply a wire clip to make the cover sit down snugly, but avoid too much pressure, which would squash the specimen out of shape or damage it. A suitable clip can be made out of a twisted paper clip. Thick objects like legs are best mounted in cavity slides (slides with depressions ground out and polished); covers must of course be large enough to cover the cavities. Put away the slides to harden in a warm place (above a radiator or stove, or in a heated airing cupboard). When the balsam is quite hard, any excess which has exuded can be scraped off the slide with a knife, and the slide can then be cleaned with xylene. Finally, the slides should be labelled fully and clearly.

The above are general instructions for mounting in balsam. Special notes for particular organs follow.

Sternites. Mount sternites of a typical segment, *e.g.,* A5, of a worker, queen, and drone, for comparison.

Tergite. Mount the tergite of A7, to show the canal of the scent gland.

Spiracles. Mount part of an abdominal tergite to include a spiracle. The propodeal spiracle should also be mounted, a small part of the propodeum being cut out, including the spiracle.

Legs. The fore-legs should be mounted to show the antenna cleaner, one of the legs being flexed and the other extended. Two hind legs should be mounted side by side, one with the inner surface upwards, the other with the outer surface upwards. The hind legs will be easier to mount in a cavity slide.

Foot. Cut off the feet of two legs, including the adjoining segment of the tarsus, and mount side by side, to show the dorsal and ventral surfaces. The claws take up the position shown in Plate 6, though they are sometimes illustrated differently.

Proboscis. Remove the proboscis of a freshly killed bee as directed in Chapter 10, section 1. Under the dissecting microscope, use needles to tear the membrane, thus allowing the proboscis to lie flat with its members in the position shown in Plate 3. Do this in a watch glass, under water. Use a sable brush to clear away pollen grains, dust, etc., which may adhere to the specimen. There is no need to bleach or macerate. Spread out the specimen on a micro-scope slide arranging it as in Plate 3. Now lay another slide on top of the specimen, and bind the two slides together with thread. Immerse the sandwich in 95 per cent. alcohol for two or three days; this will stiffen the specimen, which will retain its position throughout the remainder of the preparation. Take the proboscis out from between the slides, pass it through cellosolve, clove oil, and xylene; then mount in balsam under pressure.

Antenna. Take the antennae from a drone, partially bleach them, and macerate. Finish as directed in the general schedule. The sense organs can be seen as illustrated in Fig. 28.

The sting apparatus. Take a sting, preferably from a black bee, in which the plates will be conspicuously coloured. Do not bleach. Macerate in potash, wash, and clean up thoroughly with needles and brush, in water, under the dissecting microscope. Remove the proctiger, or it will be impossible to flatten the apparatus. Now dehydrate, clear, wash in xylene, and mount in balsam, under pressure.

This preparation shows the plates and other sclerotized parts. It is instructive also to mount stings in glycerine, showing the muscles and glands from different aspects: dorsal, ventral, and also lateral, the last with the proctiger still attached and the shaft protracted. Cavity slides will be necessary, the sting being rather a thick object; or these specimens can be displayed in a micro-jar. Mounting in glycerine is described in the next section.

Cornea. The cornea can be obtained as follows: slice off the flattest part of the compound eye, put it into water in a watch glass, and, under the dissecting microscope, scrape out the pigmented tissue (the ommatidia) with knife and brush. Rinse in water, then in alcohol, and allow the specimen to dry. If it were to be mounted in balsam, detail would become invisible and experiments impossible. It is therefore to be mounted dry, in air. Place the clean cornea on a perfectly clean slide, and cover it with a small cover glass, putting a clip on the whole. Now just touch the edge of the cover in three or four places with very small drops of balsam, just enough to prevent the cover from moving, but not preventing ventilation. *Sealed* mounts in air are not satisfactory: the cover becomes etched and cloudy. The lenses of the cornea can be demonstrated as follows: place the slide on the stage of the microscope under a 4 mm. objective. In front of the lamp place a bold transparency (positive or negative portrait, for example) or a card with a simple design ('V', for example) cut out of it. Lower the substage condenser below its proper position of focus, and rack up the microscope above the level of sharp focus; the aerial images formed by the corneal lenses will then come into focus.

Proventriculus. Cut off a proventriculus, slit it up one side, and lay it out flat, as in Fig. 16C. Tie between slides, as was done with the proboscis, and fix for an hour in 70 per cent. alcohol. Take out, and stain in a one per cent. solution in water of chlorazol fast pink. Rinse in water, dehydrate, etc., and mount in balsam.

Another useful preparation of the proventriculus can be made by cutting off enough of its head to show the lips and pouches as they are drawn in Fig. 16A. After fixing and washing, this is best mounted in glycerine, as described in the next section, in a cavity slide.

Hypopharynx. Dissect out from a head, macerate, but do not bleach. Mount flat under pressure.

Wings. The wings, with the articular sclerites, need neither bleaching nor maceration, but they should be dehydrated in cellosolve. Mount flat, under pressure, in pairs (fore and hind).

(2) The soft internal organs

These require a variety of treatments, and only a few are worth preserving as 'whole mounts'.

Ovarioles with developing eggs. Under water, tease out two or three complete ovarioles, very gently, with needles. This really amounts to breaking the fine tracheae and tracheoles which tie the ovarioles together. Immerse the ovarioles in 70 per cent. alcohol for half an hour. While they are fixing, prepare a cavity slide and cover glass. Using a turntable and a sable brush (No. 3), lay a ring of gold size round the cavity and outside it, and paint the edge of the cover glass with a ring of identical diameter, so that when the cover is applied to the slide the two rings will meet. Fill the cavity with glycerine, *overfill* it, so that the glycerine is piled up above the level of the surface of the slide. Transfer the ovarioles to the glycerine in the cavity, and cover. Excess glycerine will be forced out between the cover and the slide. Now use a needle to press the edges of the cover into contact with the slide, thus pressing the gold size rings together. Work round the edge of the cover, and examine the work under the dissecting microscope to make sure that there is good, complete union between the two rings. On no account apply pressure to the centre of the cover, or, when pressure is relaxed, air bubbles will be drawn into the mount. When the seal is complete and satisfactory, wash off the exuded glycerine under the tap, and finally rinse in distilled water; then allow the slide to dry completely. When it is quite dry the slide is ready for its final sealing. Transparent nail lacquer, asphalte varnish, or good shellac varnish, may be used. Run rings on to the edge of the cover, allowing each to harden before applying the next. Several coats, at intervals of a few days, should be applied, until the seal is thick and strong. An even stronger finish is provided by a single thick ring of Araldite, a method described by the author elsewhere (1960*b*). This preparation is suitable for examination by strong top lighting, not by transmitted light.

Glands. All the glands can be mounted by a very simple method. After dissecting out portions of the required gland (avoid large masses), they can be mounted without preliminary treatment in lactophenol (phenol crystals 20 gm., lactic acid 16 cc., glycerine 31 cc., water 20 cc., trypan blue 0·05 per cent. (see Dade, 1960*b*). The material is fixed, stained, and mounted in one operation. Heat must not be used. The fluid must not exude beyond the edges of the cover glass, or the mount cannot be sealed. Seal with nail lacquer, many coats (*no other* cement is safe), or one coat followed by Araldite.

Tracheae. Their structure, with spiral thickenings (taenidia), can be seen in temporary mounts in water. Permanent mounts can be made in balsam, after fixing in 70 per cent. alcohol, and staining in one per cent. chlorazol fast pink in water for two or three hours. They must then be rinsed quickly and dehydrated, cleared, and mounted in balsam.

Whole larva. Fix a one-day-old larva in 70 per cent. alcohol to which has been added 5 per cent. of acetic acid. Leave it in the fixative for two hours, then wash out thoroughly with 70 per cent. alcohol until there is no smell of acetic acid. Stain in borax carmine or erythrosin. Wash, dehydrate, etc., and mount in balsam in a cavity slide. These small larvae become quite transparent, and their viscera can be seen.

Spermatozoa. Obtain spermatozoa from a mature drone or from the spermatheca of a queen. Put some of the sperm in a drop of 0·15 per cent. salt solution in a watch glass or on a slide, and after teasing out put them in a warm place until they are active. Prepare a grease-free slide by rubbing it with undiluted detergent (Teepol, or moist Tide) and then wash off the detergent thoroughly under the tap, finally rinsing with distilled water; set it aside to dry. (In fact, several slides should be so prepared.) Take a drop of the suspension of spermatozoa on a glass rod, or in the pipette, and spread it in the middle of one of the cleaned slides. Allow the water to dry off, but do not apply heat. When the film is dry, it is ready to stain. Carbol fuchsin, as used for bacterial smears, is a suitable stain, but should be diluted with an equal volume of water. Flood the slides with the stain and warm it very gently until a curl of steam is seen to rise; it must not be boiled. Now set the slide aside for about half an hour, watching it to see that it does not become dry, adding more stain if necessary. Wash under the tap, and then, with the tap still running, rinse the slide with weak alcohol and *instantly* wash off the alcohol under the running tap. Continued exposure to the alcohol will remove all the stain. Finally, rinse in distilled water and allow to dry. When the slide is dry, apply a small drop of balsam and cover the preparation with a warm cover glass. Warm the slide gently to make the cover settle down.

APPENDIX III

PLURAL FORMS AND DERIVATIONS OF ANATOMICAL TERMS

Knowing the derivation of words from their Latin and Greek roots helps us to remember their meaning, besides being a matter of interest. In this list the correct plural terminations of nouns are shown following the singular forms, except when the terms are fully anglicized and therefore simply add 's' in the plural. In brackets the Latin or Greek roots are given, except where the term and/or its meaning are the same, and then only the language, Latin (L.) or Greek (Gk.) is indicated. Many of the Latin root words are themselves of Greek origin.

abdomen, -mina (L.)
acinus, -ni (L., a berry)
antenna, -ae (L., a sail yard)
anterior (from L., *ante*, before; referring to the fore part)
anus, -ni (L.)
aorta, -ae (L.)
apodeme (Gk., *apo-*, apart from, and *demas*, the frame of the body)
arcus, -ci (L., a bow or arch)
arolium, -ia (etymology obscure; perhaps from Gk. *aro*, to plough, and *leios*, smooth)
arthropod, -da (Gk., *arthron*, a joint, and *pous*, a foot; hence 'with jointed legs')
atrium, -ria (L., a court before a house)
auricle (L., the lobe of the ear)
basalare (L., *basis*, base, and *ala*, a wing)
basisternite (L., *basis*, base, and sternite, *q.v.*)
basitarsus, -rsi (L., *basis*, base, and tarsus, *q.v.*)
bursa, -ae (L., a pouch or purse)
cardo, -dines (L., a hinge)
cephalic (Gk., *kephale*, the head)
cervix, -ices (L., the neck)
chemotaxis (Gk., *chimia* (from Arabic alkimia) and *taxis*, arrangement; movement in relation to a chemical substance)
chiasma, -ata (Gk., *chiasein*, to cross, like the Gk. letter χ (chi))
chitin (Gk., *chiton*, a tunic)

cibarium, -ia (L., food)

clypeus, -pei (L., a shield)

corbicula, -ae (L., a basket)

corna, -nua (L., a horn)

cornea, -ae (L., *corneus*, horny)

corpus, -pora (L., a body)

corpus allatum, -pora -ta (L., *corpus*, a body, and *adferor*, to move towards; hence moved (during development) to their final position)

corpus cardiacum, -pora -iaca (L., *corpus*, and Gk. *kardia*, the heart; hence bodies near the heart)

corpus pedunculatum, -pora -ta (L., *corpus*, and *pedunculus*, a little foot or stalk of a flower; hence stalked bodies)

coxa, -ae (L., the hip)

cuticle (L., *cuticula*, diminutive of *cutis*, skin)

dermis (Gk., *derma*, skin)

deuto- (Gk., *deuteros*, second)

deutocerebrum (Gk., *deuteros*, and L., *cerebrum*, the brain)

diploid (Gk., *diploos*, double, and *-oeides*, having the form of)

distal (L., *distans*, distant)

dorsal (L., *dorsum*, the back; referring to the back of an animal)

ecdysis, -ses (Gk., *ekduo*, to strip or moult)

endocrine (Gk., *endon*, within, and *krino*, to separate; thus secreting *into* the body)

endophallus, -lli (Gk., *endon,* within, and *phallos*, penis)

endosternite (Gk., *endon* and sternite, *q.v.*)

epidermis (Gk., *epi-*, over or upon, and *dermis, q.v.*)

epipharynx, -nges (Gk., *epi-*, and *pharynx, q.v.*)

episternite (Gk., *epi-*, and sternite, *q.v.*)

epithelium, -ia (Gk., *epi-*, and *thele*, a nipple; extended meaning to a thin layer of tissue)

femur, -mora (L., the thigh)

fibula, -ae (L., a clasp)

flabellum, -lla (L., a little fan)

flagellum, -lla (L., a little whip)

foramen, -mina (L., a hole)

fossa, -ae (L., a ditch or trench)

frons, -ntes (L., the brow or forehead)

furca, -ae (L., a fork)

furcula, -ae (L., a little fork)

galea, -ae (L., a helmet; in some insects the galea resembles a Greek helmet)

ganglion, -ia (Gk., a tumour or swelling)

gena, -ae (L., a cheek)
glossa, -ae (L., tongue)
gynandromorph (Gk., *gyne*, a woman, *aner*, *andr*(*o*), a man, and
  morphe, shape or form)
hamulus, -li (L., a small hook)
haploid (Gk., *haploos*, one, single, and *-oeides*, having the form of)
hymenopteron, -ra (Gk., *hymen*, a membrane, and *pteryx*, a wing;
  thus, with membranous wings)
hypodermis (Gk., *hypo*, below, and dermis, *q.v.*)
hypopharynx, -nges (Gk., *hypo*, and pharynx, *q.v.*)
ileum, -lei (L., part of the small intestine)
imago, -gines (L., image or portrait)
instar (L., bigness; hence a stage of growth)
labellum, -lla (L., a little lip)
labium, -ia (L., the lower lip)
labrum, -ra (L., the upper lip)
lacinia, -ae (L., a flap or fringe)
larva, -ae (L., a disguise or ghost)
lateral (L., *lateralis*, pertaining to the side)
ligula, -ae (L., a small tongue)
lorum, -ra (L., a thong or leash)
lumen, -mina (L., light; extended to mean a cavity from its appearance in a microscope section)
mandible (from L., *mando*, to grind with the teeth; of insects, the
  upper jaw)
maxilla, -ae (L., a jaw; of insects, the *lower* jaw)
mentum, -ta (L., the chin)
mesenteron (Gk., *mesos*, middle, and *enteron*, the intestine)
mesothorax, -aces (Gk., *mesos*, middle, and thorax, *q.v.*)
metamorphosis, -ses (Gk., *meta*, change, and *morphe*, form or shape)
metathorax, -aces (Gk., *meta* and thorax, *q.v.*)
micropyle (Gk., *mikros*, small, and *pyle*, a gate or opening)
notum, -ta (Gk., *noton*, the back)
occiput (L., the back of the head)
ocellus, -lli (L., a little eye)
oenocyte (Gk., *oinos*, wine, and *kytos*, a vessel (cell); the reference
  is to the colour of the cell)
oesophagus, -gi (Gk., oisophagus)
ommatidium, -ia (Gk., *ommatidion*, a little eye)
oogenesis (Gk., *oon*, egg, and *genesis*, origin)
ostium, -ia (L., a door)
ovariole (L., a little ovary)
ovary (from L., *ovum*, an egg)

oviduct (L., *ovum* and *ductus*, a pipe or tube)

ovum, -va (L., *ovum*, egg)

paraglossa, -ae (Gk., *para-*, by the side of, and glossa, *q.v.*)

parthenogenesis (Gk., *parthenos*, a virgin, and *genesis*, origin)

pecten, -tines (L., a comb or rake) = rastellum

peduncle (L., *pedunculus*, a flower stalk) = petiole in insects

petiole (L., *petiolus*, a fruit stalk) now, in botany, a *leaf* stalk

phallotreme (Gk., *phallos*, penis, and *trema*, orfice)

pharynx, -nges (Gk., the throat or gullet)

phragma, -ae (Gk., a fence)

planta, -ae (L., the sole of the foot)

pleurite (Gk., *pleura*, a rib)

pneumophysis, -ses (Gk., *pneuma*, air, and *physis*, the nature of a growth; thus a growth containing air) = corna

postcerebral (L., *post-*, behind, and cerebrum, *q.v.*)

posterior (L., *post-*, behind) pertaining to the rear part of the body

postmentum, -ta (L., *post-*, and mentum, *q.v.*)

prementum, -ta (L., *pre-*, before, and mentum, *q.v.*)

prepupa, -ae (L., *pre-*, and pupa, *q.v.*); thus preceding the pupa

pretarsus, -si (L., *pre-*, and tarsus, *q.v.*); the foot of insects

proctiger, -ra (Gk., *proktos*, the anus, and *gero*, bear); the terminal abdominal segment bearing the anus

proctodeum, -ea (Gk., *proktos*, the anus, and *-odes*, of the nature of); the embryonic hind-gut

propodeum, -ea (L., *pro-*, in front of, and Gk., *pous*, a foot (stalk)); thus in front of the stalk (petiole)

prothorax, -aces (L., *pro-*, in front of, and thorax, *q.v.*)

protocerebrum, -ra (Gk., *protos*, first, and cerebrum, *q.v.*)

proventriculus, -li (L., *pro-*, in front of, and ventriculus, *q.v.*)

pylorus, -ri (Gk., *pyle*, a gate or opening, and L., *os*, mouth); strictly, part of the stomach

rastellum, -lla (L., a little rake)

rectum, -ta (L., *rectus*, straight); the straight terminal part of the intestine

retina, -ae (dog L., perhaps from *rete*, a net)

retinula, -ae (L., a little retina)

rhabdom (Gk., *rhabdos*, a rod)

scape (L., *scapus*, an upright stem or shaft)

sclerite (Gk., *skleros*, hard, and *-ites*, connected with)

sclerotin (Gk., *skleros*, hard)

scolophore (Gk., *skolos*, a prickle or thorn, and *phoros*, a bearer)

scopa, -ae (L., a brush or broom)

scutum, -ta (L., a shield)

scutellum, -lla (L., a little shield)
sensillum, -lla (from L., *sentio*, discern by the senses)
seta, -ae (L., a bristle)
sinus (L., *sinus*, a hollow or bay)
sperma (Gk., seed)
spermatheca, -ae (Gk., *sperma*, and *theka*, a container)
spermatogenesis (Gk., *sperma*, and *genesis*, origin)
spermatogonium, -ia (Gk., *sperma*, and *gone*, generation)
spermatophore (Gk., *sperma*, and *phoros*, that which bears or carries)
spermatozoon, -zoa (Gk., *sperma*, and *zoon*, an animal)
stadium, -ia (Gk., *stadios*, standing firm; hence a state or position)
stipes, -pites (L., a stake or trunk)
stomodeum, -ea (Gk., *stoma*, mouth, and *-odes*, of the nature of);
   the embryonic fore-gut
strigilis, -les (L., a curry-comb or scraper)
subalare (L., *sub-*, under, and *ala*, a wing)
suboesophageal (L., *sub-*, under, and oesophagus, *q.v.*)
sulcus, -ci (L., a groove or furrow)
suture (L., *sutura*, a seam)
taenidium, -ia (L., *taenia*, a ribbon, and diminutive suffix)
tarsus, -si (Gk., *tarsos*, the sole of the foot)
tarsomere (Gk., *tarsos*, and *meris*, a part)
tegula, -ae (L., a tile)
tentorium, -ia (L., a tent; hence, a tent-pole)
tergite (L., *tergum*, back, and *-ites*, of the nature of)
testis, -tes (L., testicle)
thorax, -aces (Gk., *thorax*, the breast or torso)
tibia, -ae (L., the shin bone)
trachea, -ae (L., the windpipe)
tracheole (L., *trachea* and *-ole*, diminutive suffix)
tritocerebrum, -ra (Gk., *tritos*, third, and cerebrum, *q.v.*)
trochanter (from Gk., *trecho*, to run); in vertebrates, a protuberance
   of the thigh bone)
unguis, -ues (L., a claw)
vagina, -ae (L., a scabbard or sheath)
vas deferens, -sa -ntia (L., *vas*, a vessel, and *defero*, convey)
ventral (L., *venter*, the belly); pertaining to the underside of an
   animal
ventriculus, -li (L., *venter*, and *-culus*, diminutive suffix); the mid-gut
   or stomach of insects
vertex, -tices (L., the crown of the head)
viscera (L., entrails)

FURTHER READING

In a book of this kind brief suggestions for further reading are probably more appropriate than a long formal bibliography. For over fifty years R. E. Snodgrass has been our mentor. His first book on the subject appeared in 1910, and an extended treatment of anatomy and physiology was published in 1925. These books were superseded by the great monograph of 1956, *The Anatomy of the Honeybee* (London, Constable, and Ithaca, N.Y., Comstock), to which all will turn for details of bee anatomy, and it must therefore be the first and most important title in our list. It includes a very full bibliography up to the date of its publication. Other works by Snodgrass will be noticed below.

J. A. Nelson was responsible for two publications which are still standard in the field of embryology and the anatomy of the larva. Unfortunately, they are both difficult to acquire. *The Embryology of the Honeybee* was published in 1915 by the Princeton University Press, and is a rare book. *The Morphology of the Honeybee Larva* appeared in 1924 in the *Journal of Agricultural Research*, **28**: 1167-1213. Both should be reprinted.

Everyone interested in the anatomy of the honeybee should read the first of the two volumes of F. R. Cheshire's *Bees and Beekeeping* (1886, London; reprinted 1921). Cheshire was the pioneer, being the first to write about bee anatomy in a really scientific manner; he made many original observations, and his work must always command our respect.

Important background books are the following:

A. D. Imms (1946)
A General Textbook of Entomology; London, Methuen.

R. E. Snodgrass (1935)
The Principles of Insect Morphology; London and New York, McGraw-Hill.

V. B. Wigglesworth (1947)
The Principles of Insect Physiology; London, Methuen.

Among other valuable publications by Snodgrass are 'The thorax of insects and the articulations of their wings' in *Proc. U.S. Nat. Mus.*, **36**: 511-595, 1909; 'The morphology of the insect abdomen', *Smithsonian Pub.* No. 3219, 1933; 'The male genitalia of Hymenoptera', *Smithsonian Pub.* No. 3599, 1941; and 'The skeletomuscular mechanisms of the honeybee', *Smithsonian Misc. Coll.* No. 103, 1942.

In 1927-28 G. D. Morison published three papers in *Quart. J. micr. Sci.*, **71**: 395-463, 563-651, **72**: 511-526, on 'The muscles of the adult honeybee', with much interesting information about some of the somatic and visceral muscles and their chemistry.

Consideration of anatomy and physiology cannot be divorced from other aspects of bee science, particularly that of behaviour, on which key works have been written by C. G. Butler (1954, *The World of the Honeybee*; London, Collins, with a large collection of remarkably fine photographs of the activities of bees) and C. R. Ribbands (1953, *The Behaviour and Social Life of Honeybees*; London, Bee Research Association).

Mrs. D. Hodges, in her book *The Pollen Loads of the Honeybee* (1952, London, Bee Research Association), refers to the relevant anatomy, with excellent illustrations. F. N. Howes, of the Royal Botanic Gardens, Kew, has written the best book on bee forage and the relationships between bees and plants: *Plants and Bee-keeping* (1945, London, Faber & Faber).

The World of Bees, by Gilbert Nixon (Hutchinson, 1954, also available in the Grey Arrow edition, 1959), gives a comprehensive account of the natural history of the solitary bees, stingless bees, bumble bees, etc., as well as of the honeybee, and thus shows the last named in perspective among her relatives.

J. W. S. Pringle's important *Insect Flight* (1957, Cambridge University Press) has added a great deal to our knowledge of this subject. Another valuable contribution is P. T. Haskell's *Insect Sounds* (1961, London, Witherby).

The Bee Research Association is responsible for many valuable contributions to bee science. Some of its publications have already been mentioned (books by Hodges and Ribbands). The *Bee World*, a monthly journal, contains original papers on all aspects of apiology and apiculture, and includes *Apicultural Abstracts*, with notices and abstracts of all publications in all parts of the world, thus supplementing the bibliographies in key books. '*A.A.*' is quite indispensible; it is also issued separately, and is available on index cards as well. Publication began in 1950. Another useful reference book published by the B.R.A. is the *Dictionary of Beekeeping Terms*, with the equivalents in all the principal European languages; it includes all the anatomical and other scientific terms as well as those of apiculture.

A few papers, most of them in English, to which specific reference has been made in the text are:

Bailey, L. (1952). The action of the proventriculus of the worker honeybee, *Apis mellifera* L. *J. exp. Biol.*, **29**: 310-327

Bailey, L. (1954). The respiratory currents in the tracheal system of the adult honeybee. Ibid., **31**: 589-593

Betts, A. D. (1923). Practical bee anatomy. Benson, Oxon., the Apis Club

Boxall, F. F. H. (1956). Preparation of sections by the paraffin wax process. *J. Quekett micr. Cl.*, ser. 4, **4**: 229-239

Dade, H. A. (1949). The laboratory diagnosis of honeybee diseases. Monogr. 4, *Quekett micr. Cl.*, London, 1949. (Obtainable from B.R.A.)

Dade, H. A. (1951). Amateur construction of a dissecting microscope. *Bee World*, **32**: 34-26. (Reprints obtainable from B.R.A.)

Dade, H. A. (1960a). A featherweight prismatic dissecting microscope. *J. Quekett micr. Cl.*, ser. 4, **5**: 255-256

Dade, H. A. (1960b). On mounting in fluid media, with special reference to lactophenol. Ibid., ser. 4, **5**: 308-317

Mackensen, O. and Roberts, W. C. (1948). Manual for the artificial insemination of queen bees. ET-250, *U.S. Dept. Agric., Bur. Ent. and Plt. Quarantine*

Milns, N. (1960). A cheap erecting microscope body. *J. Quekett micr. Cl.*, ser. 4, **5**: 257

Ruttner, F. (1956). The mating of the honeybee. *Bee World*, **37**: 2-14, 23-24

Ruttner, F. (1959). Der Vorgang der Paarung bei der Honigbiene. *Bienenvater* **80**: 36-39, 73-77, 99-104. (Abstr. (1961) in *Apic. Abstr.* No. 119/61)

Woods, E. F. (1956). Queen piping. *Bee World*, **37**: 185-195, 216-219

Woods, E. F. (1959). Electronic prediction of swarming in bees. *Nature, Lond.*, **184**: 842-844.

(Members of the Bee Research Association are able to borrow from the Association's extensive collection of books, journals and reprints, including many English translations of publications in other languages.)

INDEX

Part I and Appendices I and II are fully indexed. Chapters 10 to 15 of Part II are indexed in general terms only, and in the order of the work, under 'Dissection': references to organs will be found in the obvious places.

153

I

ails of the

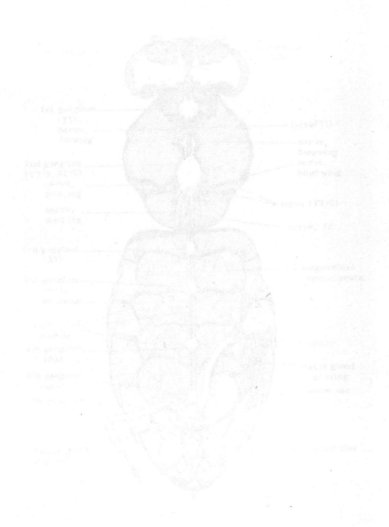

c

d

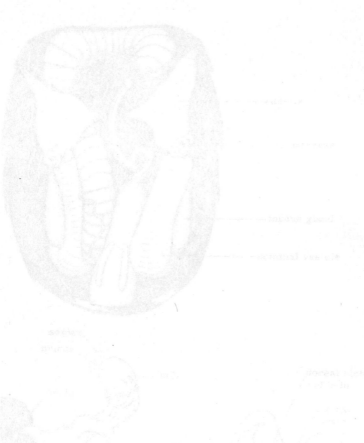

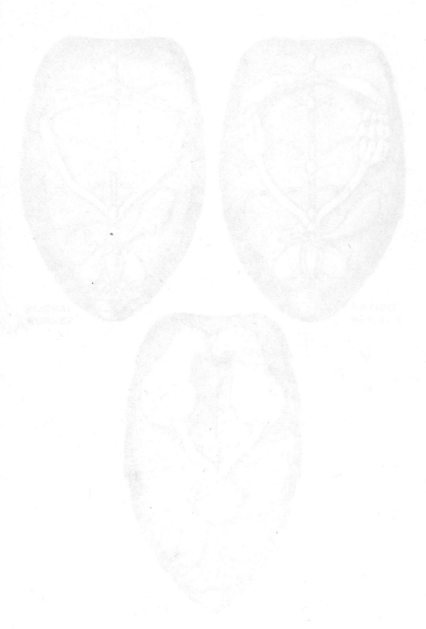

A10

side.
ials.
eor-